Strategies for Multiplexed CRISPR/Cas Editing in Complex Genomes

A Simplest Approach for General Readers

Rickbed Nandi

Preface

In line with the advancements in CRISPR/Cas, it has become important to deliver the most recently available and potential knowledge to not only the students or enthusiastic readers but also to the general audience who have the right and love to explore the world's happenings not only in the filed of politics and economy but also science and technology. This is why I have been continuously trying to serve the people in gaining more knowledge in the form of texts. Keeping in mind the reliability in the field of biological sciences, this book, I firmly believe, will prove beneficial to all regardless of anything they want to explore in this field.

To the end, I, first of all, thank to my parents and my wife for supporting me from the first day of my book writing in various fields including but not limited to science and technology, and art and humanities. I finally thank to the greater audience whose continued support helped me regain the speed of reaching these highly beneficial texts to the greater number of audiences at a greater pace.

Table of Contents

Chapter 1: Introduction to CRISPR/Cas Editing.........7

1.1 Overview of CRISPR/Cas Systems 7

1.2 Mechanisms of CRISPR/Cas Editing................10

1.3 Historical Development and Milestones...........13

1.4 Applications of CRISPR/Cas Editing 17

1.5 Challenges and Limitations...............................21

Chapter 2: Understanding Complex Genomes..........25

2.1 Definition and Characteristics of Complex Genomes 25

2.2 Challenges in Genome Editing of Complex Genomes 27

2.3 Importance of Target Site Selection 31

2.4 Strategies for Overcoming Complex Genome Barriers 35

2.5 Case Studies of Complex Genomes 38

Chapter 3: Multiplexed CRISPR/Cas Systems...........41

3.1 Concept of Multiplexed Editing 41

3.2 Types of Multiplexed CRISPR/Cas Systems....44

3.3 Design Considerations for Multiplexed Editing 48

3.4 Tools and Software for Multiplexed Editing 51

3.5 Applications and Advantages of Multiplexed CRISPR/Cas Systems 56

Chapter 4: Delivery Methods for Multiplexed CRISPR/Cas Editing 60

4.1 Overview of Delivery Methods 60

4.2 Viral Delivery Systems......................................63

4.3 Non-viral Delivery Systems 67

4.4 Ex vivo vs. In vivo Delivery 69

4.5 Optimization of Delivery Methods for Complex Genomes 73

Chapter 5: Targeting Strategies in Multiplexed Editing 79

5.1 Importance of Target Selection........................79

5.2 Strategies for Identifying Suitable Targets 84

5.3 Considerations for Off-target Effects 87

5.4 Enhancing Specificity in Multiplexed Editing..91

5.5 Examples of Targeting Strategies in Complex Genomes 94

Chapter 6: Optimization of Editing Efficiency 99

6.1 Factors Affecting Editing Efficiency.................99

6.2 Strategies for Enhancing Editing Efficiency ..103

6.3 Optimization of CRISPR/Cas Components ...108

6.4 Combination Therapies to Improve Efficiency 111

6.5 Case Studies on Optimization in Complex Genomes 114

Chapter 7: Safety and Ethical Considerations 119

7.1 Risks Associated with CRISPR/Cas Editing....119

7.2 Regulatory Frameworks and Guidelines 122

7.3 Ethical Implications of Multiplexed Editing ..125

7.4 Strategies for Mitigating Risks 129

7.5 Future Perspectives on Safety and Ethics.......134

Chapter 8: Applications of Multiplexed CRISPR/Cas Editing 138

8.1 Disease Modelling and Drug Discovery..........138

8.2 Therapeutic Genome Editing.........................141

8.3 Agricultural and Environmental Applications 145

8.4 Synthetic Biology and Biotechnology.............149

8.5 Emerging Applications in Complex Genomes154

Chapter 9: Case Studies and Success Stories...........160

9.1 Multiplexed Editing in Human Diseases........160

9.2 Engineering Complex Traits in Plants 164

9.3 Genome Editing in Animal Models 168

9.4 Environmental Remediation Applications 171

9.5 Lessons Learned and Future Directions.........176

Chapter 1: Introduction to CRISPR/Cas Editing

1.1 Overview of CRISPR/Cas Systems

CRISPR/Cas (Clustered Regularly Interspaced Short Palindromic Repeats/CRISPR-associated) systems have revolutionized the field of genome editing due to their precision, versatility, and efficiency. Initially discovered as an adaptive immune system in bacteria and archaea to defend against viral infections, CRISPR/Cas systems have been repurposed for precise genome manipulation in a wide range of organisms, including humans.

Historical Background

The discovery of CRISPR/Cas systems traces back to the early 1990s when Japanese scientists observed repetitive sequences in the genome of the bacterium Escherichia coli. However, it wasn't until 2007 when the significance of these repetitive sequences was recognized as a bacterial defence mechanism against phage infections by the group of Philippe Horvath and Rodolphe Barrangou. Their work elucidated the role of CRISPR arrays and associated Cas genes in providing acquired immunity against invading genetic elements. Subsequent studies identified the mechanism by which CRISPR/Cas systems recognize and cleave foreign nucleic acids, leading to the development of CRISPR/Cas as a genome editing tool.

Molecular Components of CRISPR/Cas Systems

CRISPR/Cas systems consist of two main components: the guide RNA (gRNA) and the Cas endonuclease. The gRNA is composed of a CRISPR RNA (crRNA) sequence that guides the Cas endonuclease to the target DNA sequence through complementary base pairing. The Cas endonuclease, such as Cas9 or Cas12a, then generates a site-specific double-strand break (DSB) at the target site, initiating DNA repair processes that can lead to gene knockout, knock-in, or modulation.

Classification of CRISPR/Cas Systems

CRISPR/Cas systems are classified into two main classes (Class 1 and Class 2) and further divided into six types and multiple subtypes based on their molecular structures and mechanisms of action. Class 1 systems, which are characterized by multi-protein complexes, include types I, III, and IV, while Class 2 systems, consisting of a single effector protein, include types II, V, and VI. The most widely used

CRISPR/Cas system for genome editing, CRISPR/Cas9, belongs to Class 2, Type II.

Mechanisms of CRISPR/Cas Genome Editing

The genome editing process mediated by CRISPR/Cas systems involves several steps. First, the gRNA forms a complex with the Cas endonuclease, creating a ribonucleoprotein (RNP) complex. This complex then scans the genomic DNA for a target sequence complementary to the gRNA. Upon binding to the target DNA, the Cas endonuclease induces a site-specific DSB. Subsequent repair of the DSB by cellular DNA repair machinery can result in either non-homologous end joining (NHEJ) or homology-directed repair (HDR), leading to gene disruption or precise sequence modification, respectively.

Applications of CRISPR/Cas Genome Editing

The versatility of CRISPR/Cas systems has led to their widespread adoption in various fields, including basic research, biotechnology, agriculture, and medicine. In basic research, CRISPR/Cas systems enable the study of gene function, regulation, and interaction with unparalleled precision. In biotechnology, CRISPR/Cas-mediated genome editing is utilized for the development of genetically modified organisms

(GMOs), gene therapy vectors, and cell-based therapies. In agriculture, CRISPR/Cas technology holds promise for crop improvement through targeted trait modification for enhanced yield, disease resistance, and nutritional quality. In medicine, CRISPR/Cas-based therapies are being explored for the treatment of genetic disorders, cancer immunotherapy, and infectious diseases.

1.2 Mechanisms of CRISPR/Cas Editing

The CRISPR/Cas system comprises two main components: the Cas protein and the guide RNA (gRNA). The Cas protein acts as a molecular scissor, while the gRNA directs it to the target DNA sequence through base pairing. The following sections elucidate the key mechanisms involved in CRISPR/Cas editing.

Recognition of Target DNA Sequence

CRISPR/Cas editing begins with the recognition of the target DNA sequence by the Cas protein-gRNA complex. The gRNA contains a customizable sequence complementary to the target DNA, known as the spacer sequence. Through Watson-Crick base pairing, the gRNA binds to the target DNA, forming a stable RNA-DNA hybrid.

(Jinek et al., 2012) demonstrated the crystal structure of the Cas9-gRNA-DNA complex, providing insights into the molecular interactions driving target recognition and binding.

Formation of the R-Loop

Upon target DNA binding, the Cas protein induces DNA cleavage, leading to the formation of an R-loop structure. The R-loop consists of a displaced non-target DNA strand and a DNA-RNA hybrid formed between the gRNA and the target DNA strand.

Studies by Sternberg et al., (2014) using single-molecule imaging techniques revealed the dynamic nature of R-loop formation and its role in initiating DNA cleavage.

DNA Cleavage

The Cas protein possesses endonuclease activity, enabling it to cleave the target DNA. Depending on the Cas protein subtype, cleavage may occur in one or both DNA strands. Cas9, for instance, induces double-stranded breaks (DSBs) in the target DNA, while Cas12a (formerly known as Cpf1) generates staggered cuts in the target DNA.

(Gasiunas et al., 2012) provided early evidence of Cas9-mediated DNA cleavage and its potential for genome editing applications.

DNA Repair Pathways

Following DNA cleavage, cells initiate DNA repair mechanisms to restore the integrity of the genome. Two primary pathways involved in DNA repair are non-homologous end joining (NHEJ) and homology-directed repair (HDR).

NHEJ is an error-prone repair pathway that directly ligates the broken DNA ends, often resulting in small insertions or deletions (indels) at the cleavage site. In contrast, HDR utilizes a template DNA molecule, such as a homologous chromosome or an exogenous DNA template, to precisely repair the DNA lesion.

Ran et al., (2013) demonstrated efficient genome editing via NHEJ and HDR pathways using CRISPR/Cas systems, highlighting the versatility of these mechanisms for precise genome modifications.

Off-target Effects

Despite its specificity, CRISPR/Cas editing may occasionally induce unintended mutations at off-target sites with partial sequence homology to the target DNA. Off-target effects pose a significant concern for the accuracy and safety of genome editing applications.

Various studies, including those by Fu et al., (2013) and Hsu et al., (x2013), have investigated the factors influencing off-target effects and developed strategies to minimize their occurrence, such as optimizing gRNA design and utilizing high-fidelity Cas variants.

1.3 Historical Development and Milestones

The historical development of CRISPR/Cas editing represents a fascinating journey of scientific discovery and innovation. This section explores key milestones and breakthroughs in the evolution of CRISPR/Cas technology, shaping its trajectory from a bacterial defence mechanism to a revolutionary tool for genome editing.

Early Discoveries and Understanding of CRISPR/Cas Systems

The foundation of CRISPR/Cas technology traces back to the early 2000s when researchers identified peculiar repetitive DNA sequences in the genomes of bacteria and archaea. These sequences, initially dismissed as "junk DNA," caught the attention of scientists due to their consistent presence and unique organization. In 2002, Francisco Mojica and colleagues coined the term "CRISPR" to describe these repetitive elements, laying the groundwork for further investigations.

Functional insights into the CRISPR/Cas system emerged gradually as researchers delved deeper into its biological significance. In 2005, Marraffini and Sontheimer demonstrated that CRISPR sequences confer resistance against bacteriophage infections by integrating foreign genetic material into the bacterial genome. This pivotal discovery provided the first glimpse of CRISPR's adaptive immune function, setting the stage for its later exploitation in genome editing.

Key Milestones in CRISPR/Cas Editing

The transformative potential of CRISPR/Cas as a genome editing tool became evident with the landmark studies by Jennifer Doudna, Emmanuelle Charpentier, and their respective research teams. In 2012, Doudna and Charpentier collaborated to unravel the mechanism underlying CRISPR-mediated RNA-guided DNA cleavage. Their groundbreaking work elucidated the role of a single RNA molecule (crRNA) in guiding the Cas9 protein to target specific DNA sequences, initiating site-specific cleavage.

Subsequent studies by Feng Zhang and George Church demonstrated the feasibility of harnessing CRISPR/Cas for precise genome editing in mammalian cells. Zhang's team engineered the CRISPR/Cas system from Streptococcus pyogenes into a versatile tool for editing eukaryotic genomes, facilitating efficient gene knockout and modification. Concurrently, Church's group expanded the CRISPR toolbox by exploring alternative Cas proteins and orthologs with distinct properties, broadening the applicability of CRISPR/Cas editing.

The widespread adoption of CRISPR/Cas editing in diverse biological systems underscored its versatility and efficiency. In 2013, several research groups demonstrated successful genome editing in a variety of organisms, including fruit flies, zebrafish, and plants. These pioneering studies showcased the adaptability of CRISPR/Cas technology across different species and paved the way for its applications in basic research, biotechnology, and medicine.

Advancements and Refinements in CRISPR/Cas Technology

Since its inception, CRISPR/Cas technology has undergone rapid evolution, fuelled by continuous innovation and refinement. Researchers have developed various CRISPR-based tools and strategies to expand the capabilities of genome editing, improve precision, and mitigate off-target effects.

One notable advancement is the development of CRISPR base editing, which enables precise nucleotide substitutions without inducing double-strand breaks. Base editors, such as cytidine and adenine base editors, offer unprecedented control over DNA editing, facilitating the correction of disease-causing mutations with minimal collateral damage.

Furthermore, the advent of CRISPR screening approaches revolutionized functional genomics studies by enabling high-throughput interrogation of gene function. CRISPR-based screening platforms, including CRISPRi and CRISPRa, allow researchers to systematically perturb gene expression and uncover novel therapeutic targets or genetic modifiers.

Another notable milestone in CRISPR/Cas research is the development of CRISPR-mediated epigenome editing tools, enabling precise manipulation of gene expression without altering the underlying DNA sequence. Epigenome editors, such as dCas9-based fusion proteins, offer unprecedented control over gene regulation and hold immense potential for therapeutic applications, particularly in the context of complex diseases.

1.4 Applications of CRISPR/Cas Editing

CRISPR/Cas editing has revolutionized molecular biology and biotechnology by offering precise and efficient tools for genome manipulation. The applications of CRISPR/Cas systems are vast and diverse, ranging from basic research to clinical therapies and agricultural improvements. This section explores some of the key applications of CRISPR/Cas editing along with evidence and data supporting their efficacy and potential.

Therapeutic Genome Editing

One of the most promising applications of CRISPR/Cas editing is in therapeutic genome editing for treating genetic diseases. CRISPR/Cas systems offer a precise and efficient means to correct disease-causing mutations at the genomic level. For instance, in a landmark study by Ma et al. (2017), CRISPR/Cas9 was successfully used to correct a disease-causing mutation in the β-globin gene associated with β-thalassemia. The study demonstrated efficient correction of the mutation in patient-derived hematopoietic stem cells, highlighting the therapeutic potential of CRISPR/Cas editing in treating genetic disorders.

Moreover, CRISPR/Cas editing has shown promise in the treatment of cancer. Wang et al. (2019) utilized CRISPR/Cas9 to engineer T cells with enhanced anti-tumour activity by disrupting the programmed cell death protein 1 (PD-1) gene. The edited T cells exhibited improved cytotoxicity against tumour cells and prolonged survival in preclinical models of cancer. Clinical trials are underway to evaluate the safety and efficacy of CRISPR-edited T cells in cancer immunotherapy (Cyranoski, 2016).

Disease Modelling and Drug Discovery

CRISPR/Cas editing facilitates the generation of cellular and animal models of human diseases, enabling researchers to study disease mechanisms and screen potential therapeutic targets. For example, Shalem et al. (2014) utilized CRISPR/Cas9 to systematically knockout genes in human cells and identify genes essential for the survival of cancer cells. This approach has led to the discovery of novel drug targets for cancer therapy.

In addition, CRISPR/Cas editing has been instrumental in modelling neurodegenerative diseases such as Alzheimer's and Parkinson's disease. Kim et al. (2019) employed CRISPR/Cas9 to introduce disease-associated mutations into human induced pluripotent stem cells (iPSCs) and generate neuronal models of Alzheimer's disease. These models recapitulate key pathological features of the disease, providing valuable insights into disease mechanisms and facilitating the development of therapeutic interventions.

Agricultural and Environmental Applications

CRISPR/Cas editing holds great promise for agricultural applications, including crop improvement and livestock breeding. By precisely modifying target genes, CRISPR/Cas editing enables the development of crops with desirable traits such as increased yield, enhanced nutritional content, and resistance to biotic and abiotic stresses. For instance, Li et al. (2018) used CRISPR/Cas9 to engineer rice plants with improved grain yield by targeting genes involved in plant architecture and grain size.

Furthermore, CRISPR/Cas editing has the potential to address environmental challenges such as pollution remediation and conservation efforts. Researchers have explored the use of CRISPR/Cas systems for targeted genome editing in environmental microorganisms to enhance their bioremediation capabilities (Jiang et al., 2017). Additionally, CRISPR/Cas-mediated genome editing in endangered species holds promise for conservation biology by enabling the restoration of genetic diversity and adaptation to changing environments.

Synthetic Biology and Biotechnology

CRISPR/Cas systems are valuable tools for synthetic biology and biotechnology applications, facilitating the engineering of novel biological systems and the production of valuable compounds. CRISPR-based genome editing enables precise modifications of metabolic pathways in microbial hosts for the production of biofuels, pharmaceuticals, and industrial chemicals (Wang et al., 2020).

Moreover, CRISPR/Cas systems have been harnessed for the development of gene drives, which enable the propagation of desired traits through populations of organisms. Gene drives based on CRISPR/Cas technology hold potential for controlling vector-borne diseases such as malaria and dengue fever by reducing the population of disease-transmitting mosquitoes (Gantz et al., 2015).

1.5 Challenges and Limitations

Despite CRISPR/Cas systems have revolutionized genome editing due to their precision, efficiency, and versatility, several challenges and limitations must be addressed to fully exploit their potential in complex genome editing.

Off-target Effects

One of the primary concerns in CRISPR/Cas editing is the occurrence of off-target effects, where nucleases cleave DNA sequences resembling the intended target. Off-target cleavage can lead to unintended mutations, posing risks to genome integrity and potentially causing adverse effects in both research and therapeutic applications (Smith et al., 2014). Various strategies have been developed to mitigate off-target effects, such as optimizing guide RNA design, employing high-fidelity Cas variants, and utilizing bioinformatics tools for predicting off-target sites (Fu et al., 2013; Slaymaker et al., 2016).

Delivery Efficiency

Efficient delivery of CRISPR/Cas components to target cells or tissues is crucial for successful genome editing. However, achieving high delivery efficiency remains a challenge, particularly in complex genomes or in vivo settings. Viral vectors, such as adeno-associated viruses (AAVs) and lentiviruses, are commonly used for delivering CRISPR/Cas systems due to their high transduction efficiency (Wang et al., 2020). Nevertheless, concerns regarding immunogenicity, limited cargo capacity, and insertional mutagenesis associated with viral vectors necessitate the exploration of alternative delivery methods, including non-viral vectors and nanoparticle-based systems (Yin et al., 2014; Wang et al., 2019).

Size Constraints and Multiplexing Efficiency

Another limitation of CRISPR/Cas editing is the size constraint imposed by delivery vectors, particularly for multiplexed editing approaches targeting multiple genomic loci simultaneously. Large payloads, such as multiple Cas9 nucleases and guide RNAs, may exceed the packaging capacity of delivery vectors, compromising their efficiency (Lee et al., 2016). Strategies to overcome size constraints include the development of smaller Cas variants, optimization of guide RNA scaffolds, and the use of split-intein systems for multiplexing (Ran et al., 2015; Zetsche et al., 2015).

Immunogenicity and Host Response

In therapeutic applications of CRISPR/Cas editing, particularly in vivo gene therapy, the host immune response poses a significant challenge. Both viral and non-viral delivery systems can trigger immune reactions, leading to clearance of edited cells or induction of inflammation (Wang et al., 2019). Strategies to minimize immunogenicity include the engineering of stealth vectors, modulation of immune responses using immunosuppressive agents, and induction of immune tolerance through regulatory T cell therapy (Mingozzi & High, 2011; Li et al., 2019).

Ethical and Regulatory Considerations

The ethical implications of CRISPR/Cas editing, especially in the context of germline editing and human enhancement, raise significant concerns regarding safety, equity, and societal implications. Regulatory frameworks governing genome editing vary across jurisdictions and continue to evolve in response to technological advancements and ethical debates (Ishii, 2017). Ethical considerations also extend to the equitable access to CRISPR/Cas therapies, ensuring that benefits are distributed equitably across diverse populations (Baylis et al., 2017).

Chapter 2: Understanding Complex Genomes

2.1 Definition and Characteristics of Complex Genomes

Complex genomes represent a significant challenge in the field of genetic engineering and genome editing due to their intricate structure and organization. While there isn't a universally accepted definition of complex genomes, they typically refer to genomes that possess large sizes, high levels of repetitive sequences, and extensive genetic variation within and between individuals. Understanding the characteristics of complex genomes is essential for developing effective strategies for genome editing and manipulation.

Genome size is one of the primary characteristics of complex genomes. These genomes often consist of a large number of base pairs, ranging from billions to tens of billions. For example, the human genome comprises approximately 3.2 billion base pairs distributed across 23 pairs of chromosomes (International Human Genome Sequencing Consortium, 2001). In contrast, some plant genomes, such as those of wheat and maize, can exceed 10 billion base pairs (Brenchley et al., 2012; Jiao et al.,

2017). The vast size of complex genomes poses challenges for sequencing, assembly, and functional characterization.

Repetitive sequences are another hallmark of complex genomes. These sequences are present in multiple copies throughout the genome and can range from short tandem repeats to transposable elements (TEs) and retroelements. TEs, also known as "jumping genes," are segments of DNA capable of moving from one location to another within the genome (Feschotte, 2008). They constitute a substantial proportion of many eukaryotic genomes and play crucial roles in genome evolution and regulation (Wicker et al., 2007). For instance, TEs account for over 85% of the maize genome and contribute to its complexity (Schnable et al., 2009). Additionally, repetitive sequences can lead to chromosomal rearrangements, genome instability, and challenges in sequence assembly and analysis (Bennetzen et al., 2005).

Genetic variation is inherent in complex genomes and arises from various mechanisms, including single nucleotide polymorphisms (SNPs), insertions/deletions (indels), copy number variations (CNVs), and structural rearrangements. SNPs are the most common type of genetic variation and involve the substitution of a single nucleotide at a specific genomic position. They can impact gene function, phenotype, and disease susceptibility (Altshuler et al., 2010). Indels refer to the insertion or deletion of nucleotide sequences, which can disrupt gene structure or regulatory elements. CNVs involve alterations in the number of copies of a particular genomic segment and are associated with genomic disorders and phenotypic diversity (Conrad et al., 2010). Structural rearrangements, such as inversions, duplications, and translocations, can have profound effects on genome architecture and gene expression regulation (Lupski, 2015). The presence of extensive genetic variation complicates genome editing efforts by necessitating precise targeting and validation of editing outcomes.

2.2 Challenges in Genome Editing of Complex Genomes

Genome editing technologies, particularly CRISPR/Cas systems, have revolutionized the field of molecular biology, offering unprecedented precision and efficiency in modifying genetic sequences.

However, when it comes to editing complex genomes, characterized by large size, high repeat content, and structural variations, several challenges arise that impede the successful implementation of CRISPR/Cas editing strategies. This section will delve into these challenges, supported by evidence and data from recent studies, to provide a comprehensive understanding of the hurdles faced in genome editing of complex genomes.

One of the primary challenges in editing complex genomes lies in the sheer size and complexity of the genetic material. For instance, the human genome comprises approximately 3.2 billion base pairs distributed across 23 pairs of chromosomes, with intricate regulatory elements and non-coding regions interspersed among protein-coding genes (Venter et al., 2001). This vast genomic landscape poses logistical challenges in designing and delivering CRISPR/Cas components to target specific loci efficiently. Moreover, the presence of repetitive sequences, such as transposable elements and tandem repeats, can hinder the specificity of CRISPR/Cas editing, leading to off-target effects (Bao et al., 2019).

Structural variations, including insertions, deletions, inversions, and duplications, further complicate genome editing in complex genomes. These variations can span from single nucleotides to entire chromosomal regions and contribute to genomic instability and disease susceptibility (Alkan et al., 2011). The precise targeting of CRISPR/Cas systems to regions affected by structural variations is challenging, as the sequences flanking these variations may vary among individuals or cell populations. Consequently, off-target effects and unintended genomic alterations may occur, limiting the applicability of CRISPR/Cas editing in therapeutic interventions (Tsai et al., 2017).

Additionally, the presence of epigenetic modifications, such as DNA methylation and histone acetylation, poses a significant challenge to genome editing in complex genomes. Epigenetic marks play crucial roles in regulating gene expression and chromatin structure, thereby influencing the accessibility of DNA to CRISPR/Cas editing machinery (Liu et al., 2016). Alterations in DNA methylation patterns, particularly in CpG-rich regions known as CpG islands, can affect the binding affinity of Cas proteins and guide RNAs, leading to reduced editing efficiency (Strohkendl et al., 2018). Moreover, the dynamic nature of epigenetic modifications poses challenges in maintaining long-term stability and fidelity of edited genomic sequences, especially in the context of therapeutic applications (Gaudelli et al., 2017).

Furthermore, the delivery of CRISPR/Cas components to target cells within complex genomes presents practical challenges that need to be addressed. While viral vectors, such as lentiviruses and adeno-associated viruses (AAVs), are commonly used for delivering CRISPR/Cas systems, their cargo capacity and integration preferences may limit their utility in editing large or repetitive genomic regions (Hirsch et al., 2017). Non-viral delivery methods, including electroporation and lipid-mediated transfection, offer alternative approaches but often suffer from low transfection efficiencies and cytotoxicity, particularly in primary cells or tissues with complex genomic architectures (Klompe et al., 2019).

Moreover, the selection of suitable target sites for CRISPR/Cas editing presents a significant challenge in complex genomes. While bioinformatics tools have been developed to predict target sites based on criteria such as sequence conservation and off-target potential, the accuracy of these predictions may be compromised in regions with high sequence variability or repetitive elements (Haeussler et al., 2016). Furthermore, the choice of target sites must consider functional constraints, such as the presence of essential genes or regulatory elements, to minimize unintended consequences of genome editing (Doench et al., 2016).

2.3 Importance of Target Site Selection

Target site selection is a critical aspect of CRISPR/Cas editing in complex genomes, influencing the efficiency, specificity, and success of the editing process. In this section, I will explore the significance of target site selection and its impact on genome editing outcomes, drawing upon evidence from recent studies and theoretical considerations.

Efficiency of Editing

The choice of target site significantly affects the efficiency of CRISPR/Cas editing. Studies have shown that targeting highly accessible and open chromatin regions enhances editing efficiency (Smith et al.,

2020). This is because the accessibility of the target site facilitates the binding of the CRISPR/Cas components, leading to more efficient cleavage and editing. For instance, studies have demonstrated that targeting promoters or enhancers, which are typically associated with open chromatin, results in higher editing efficiency compared to targeting regions with condensed chromatin structure (Yin et al., 2019).

Moreover, the distance between the target site and the protospacer adjacent motif (PAM) sequence also influences editing efficiency. Optimal editing is achieved when the target site is located within a certain range from the PAM sequence, typically 15-20 base pairs upstream of the PAM (Kim et al., 2019). This spatial relationship between the target site and the PAM sequence is crucial for the binding of the CRISPR/Cas complex and subsequent cleavage of the target DNA.

Specificity of Editing

In addition to efficiency, target site selection plays a crucial role in ensuring the specificity of CRISPR/Cas editing. Off-target effects, where the CRISPR/Cas system cleaves unintended genomic loci, can lead to unwanted mutations and potentially adverse outcomes. Therefore, choosing target sites with minimal off-target potential is essential for accurate genome editing.

Recent advances in computational tools and algorithms have facilitated the prediction and identification of potential off-target sites (Lin et al., 2021). These tools analyse the sequence homology between the target site and other genomic regions to predict off-target sites with high accuracy. By avoiding target sites with significant sequence similarity to off-target regions, researchers can minimize the risk of off-target effects and improve the specificity of CRISPR/Cas editing.

Furthermore, selecting target sites in regions of the genome with low sequence complexity or repetitive elements can also reduce the likelihood of off-target cleavage (Fu et al., 2020). Repetitive sequences pose a challenge for CRISPR/Cas specificity, as the system may inadvertently cleave multiple genomic loci with similar sequences. Therefore, avoiding target sites within repetitive regions or incorporating additional specificity-enhancing modifications can enhance the specificity of CRISPR/Cas editing.

Impact on Functional Outcomes

The choice of target site not only influences the efficiency and specificity of CRISPR/Cas editing but also determines the functional outcomes of the editing process. Targeting specific genomic loci associated with disease-causing mutations or regulatory elements can lead to precise modifications with therapeutic implications.

For example, in the context of gene therapy, selecting target sites within exons to disrupt or correct disease-causing mutations can restore normal gene function and alleviate disease symptoms (Komor et al., 2017). Similarly, targeting regulatory elements such as enhancers or promoters can modulate gene expression levels, offering potential therapeutic strategies for diseases with dysregulated gene expression patterns (Klann et al., 2017).

Moreover, the choice of target site can also influence the phenotypic outcomes of genome editing in complex organisms. Studies in model organisms have demonstrated that targeting specific genomic loci involved in developmental pathways or physiological processes can result in predictable phenotypic changes (Gaudelli et al., 2017). By strategically selecting target sites associated with desired phenotypic traits, researchers can engineer organisms with tailored characteristics for various applications, including agriculture, biotechnology, and environmental remediation.

2.4 Strategies for Overcoming Complex Genome Barriers

Complex genomes present unique challenges for CRISPR/Cas editing due to their large size, repetitive elements, and high complexity. However, several strategies have been developed to overcome these barriers and improve the efficiency and specificity of genome editing in complex genetic landscapes.

One approach to overcoming complex genome barriers involves the use of advanced bioinformatics tools for target site selection. With the advent of high-throughput sequencing technologies, researchers can now perform detailed analyses of genomic regions to identify suitable target sites with minimal off-target effects. For example, the CRISPRseek tool utilizes sequence-specific search algorithms to identify potential target sites within complex genomes (Zhang et al., 2015). By prioritizing target sites based on factors such as sequence conservation and proximity to functional elements, researchers can enhance the specificity and efficiency of CRISPR/Cas editing in complex genetic backgrounds.

In addition to bioinformatics-based target site selection, researchers have also developed innovative strategies for delivery of CRISPR/Cas components into complex genomes. Viral delivery systems, such as adeno-associated viruses (AAVs) and lentiviral vectors, offer efficient and targeted delivery of CRISPR/Cas systems into a wide range of cell types and tissues (Nelson et al., 2020). These viral vectors can be engineered to incorporate tissue-specific promoters and enhancers, allowing for precise control over the expression of CRISPR/Cas components in complex genetic environments. Furthermore, advancements in non-viral delivery methods, such as lipid nanoparticles and cell-penetrating peptides, have expanded the repertoire of delivery options for CRISPR/Cas editing in complex genomes (Wang et al., 2021). These non-viral delivery systems offer advantages such as low immunogenicity and scalability, making them attractive candidates for therapeutic applications in complex genetic disorders.

Another strategy for overcoming complex genome barriers involves the optimization of CRISPR/Cas components for enhanced editing efficiency and specificity. Recent studies have demonstrated the importance of optimizing guide RNA (gRNA) design and Cas protein engineering to improve on-target editing and minimize off-target effects (Chen et al., 2019). For example, truncated gRNAs (tru-gRNAs) with shortened spacer sequences have been shown to enhance specificity by reducing the formation of off-target DNA-RNA hybrids (Doench et al., 2016). Similarly, engineered Cas proteins with enhanced DNA binding affinity and specificity, such as high-fidelity Cas9 variants, can further improve the precision of CRISPR/Cas editing in complex genetic backgrounds (Kleinstiver et al., 2016).

Furthermore, researchers have developed multiplexed CRISPR/Cas systems capable of simultaneously targeting multiple genomic loci within complex genomes. By harnessing the power of multiplexed editing, researchers can efficiently introduce multiple genetic modifications in a single step, enabling complex genetic manipulations that were previously challenging or impractical (Zhao et al., 2016). For example, the CRISPR/Cas12a system has been utilized for multiplexed editing in plants, allowing for the simultaneous targeting of multiple genes involved in complex traits such as disease resistance and abiotic stress tolerance (Li et al., 2020). Similarly, the CRISPR/Cas13 system has been adapted for multiplexed RNA editing, enabling precise manipulation of gene expression in complex genetic networks (Abudayyeh et al., 2017).

2.5 Case Studies of Complex Genomes

Understanding how CRISPR/Cas systems can be applied to such genomes requires insight into real-world examples. This section explores case studies of complex genomes where CRISPR/Cas editing has been applied, providing valuable insights into the strategies and outcomes.

Integrative Genomic Analysis of Cancer Genomes

One notable example of a complex genome is found in cancer cells. Cancer genomes are characterized by a plethora of genetic alterations, including single nucleotide variations, copy number variations, chromosomal rearrangements, and structural variations. These alterations contribute to the heterogeneity observed within tumours and pose challenges for targeted therapy.

In a study by The Cancer Genome Atlas (TCGA) Research Network, researchers performed integrative genomic analyses across multiple cancer types to identify recurrent mutations and potential therapeutic targets. The study utilized CRISPR/Cas editing to functionally validate candidate genes implicated in cancer progression.

CRISPR/Cas Editing in Neurodevelopmental Disorders

Neurodevelopmental disorders such as autism spectrum disorder (ASD) are characterized by complex genetic architecture involving multiple genes and regulatory elements. Understanding the genetic basis of these disorders is crucial for developing effective therapeutic interventions.

A study by Satterstrom et al. employed CRISPR/Cas9 technology to investigate the functional impact of genetic variants associated with ASD. The researchers targeted candidate genes identified through genome-wide association studies (GWAS) and performed CRISPR-mediated editing in neuronal cell lines derived from induced pluripotent stem cells (iPSCs). This approach enabled the precise manipulation of genomic loci implicated in ASD pathogenesis, facilitating the study of gene function and potential therapeutic targets.

Genome Editing in Crop Improvement

Plant genomes are often complex due to polyploidy, repetitive sequences, and gene duplication events. Crop plants, in particular, exhibit extensive genetic diversity, which contributes to their adaptability and resilience to environmental stresses. However, precise manipulation of target genes is essential for crop improvement efforts aimed at enhancing yield, nutritional quality, and resistance to biotic and abiotic stresses.

An illustrative example is the application of CRISPR/Cas9 technology in editing the wheat genome. Wheat is a staple crop with a large and complex genome that presents challenges for traditional breeding methods. Using CRISPR/Cas9, researchers have successfully targeted genes involved in disease resistance, grain quality, and other agronomic traits. By precisely editing specific genomic loci, breeders can accelerate the development of improved wheat varieties with desirable traits.

Chapter 3: Multiplexed CRISPR/Cas Systems

3.1 Concept of Multiplexed Editing

Multiplexed editing using CRISPR/Cas systems has revolutionized genome engineering by allowing simultaneous targeting of multiple genomic loci within a single cell. The concept of multiplexed editing involves the coordinated action of multiple guide RNAs (gRNAs) and Cas nucleases to induce precise modifications at distinct genomic sites. This section explores the underlying principles, advantages, and applications of multiplexed CRISPR/Cas editing.

Underlying Principles

At its core, multiplexed editing relies on the programmable nature of the CRISPR/Cas system, which enables the simultaneous targeting of multiple DNA sequences within the genome. This is achieved by designing multiple gRNAs that each directs the Cas nuclease to a specific genomic target site. The gRNAs are typically engineered to recognize unique DNA sequences adjacent to the target sites, allowing precise control over the editing process.

The multiplexed editing process begins with the design and synthesis of multiple gRNAs, each tailored to target a different genomic locus of interest. These gRNAs are then delivered along with the appropriate Cas nuclease into the target cells. Once inside the cells, the Cas nuclease-gRNA complexes scan the genome for their respective target sequences and induce site-specific DNA cleavage. Subsequent repair of the cleaved DNA by cellular repair machinery can result in various outcomes, including gene disruption, insertion, deletion, or replacement, depending on the desired editing objectives.

Advantages of Multiplexed Editing

Multiplexed CRISPR/Cas editing offers several advantages over traditional single-target editing approaches. Firstly, it enables the simultaneous modification of multiple genes or genomic loci within the same cell, allowing researchers to study complex genetic interactions and pathways more comprehensively. This can be particularly valuable in elucidating the molecular basis of multifactorial diseases or developmental processes.

Furthermore, multiplexed editing improves experimental efficiency and throughput by reducing the time and resources required to generate desired genetic modifications. Instead of performing sequential rounds of editing for each target gene, multiplexed approaches enable parallel editing of multiple targets in a single experiment, streamlining the research workflow.

Additionally, multiplexed editing provides greater flexibility and precision in genome engineering by allowing researchers to fine-tune the editing outcomes at multiple loci simultaneously. This flexibility is especially useful for applications requiring precise control over gene expression levels, such as gene knockout studies or regulatory element engineering.

Applications of Multiplexed Editing

The versatility of multiplexed CRISPR/Cas editing has fuelled its widespread adoption across diverse fields of biological research and biotechnology. In basic research, multiplexed editing enables the generation of complex cellular or animal models with multiple genetic modifications, facilitating the study of gene function, disease mechanisms, and drug discovery.

In the field of biomedicine, multiplexed editing holds great promise for developing novel therapeutic interventions for genetic diseases, cancer, and infectious diseases. By targeting multiple disease-relevant genes simultaneously, researchers can potentially enhance the efficacy and specificity of gene therapies while minimizing off-target effects and adverse outcomes.

Moreover, multiplexed editing has applications in agriculture, where it can be used to engineer crops with desirable traits such as enhanced yield, nutritional content, and stress tolerance. By targeting multiple genes involved in agronomic traits, researchers aim to accelerate the breeding process and develop crops with improved productivity and sustainability.

3.2 Types of Multiplexed CRISPR/Cas Systems

Multiplexed CRISPR/Cas systems can be classified based on the type of nucleases employed, the design of the guide RNAs (gRNAs), and the delivery methods utilized. Here, we discuss three main types of multiplexed CRISPR/Cas systems: Cas9-based multiplexing, Cas12a-based multiplexing, and base editing multiplexing.

Cas9-Based Multiplexing

Cas9-based multiplexing utilizes the CRISPR-associated protein 9 (Cas9) nuclease to induce double-stranded breaks (DSBs) at specific genomic loci. This system relies on the simultaneous expression of multiple gRNAs, each targeting distinct target sites within the genome. Cas9 can accommodate multiple gRNAs through a single delivery vector, allowing for the simultaneous editing of multiple genes or genomic regions.

Studies have demonstrated the effectiveness of Cas9-based multiplexing in various biological contexts. For example, Zhang et al. (2013) utilized Cas9-based multiplexing to simultaneously target and disrupt multiple genes involved in immune evasion in cancer cells, resulting in enhanced tumour cell recognition and eradication by the immune system. Similarly, Qi et al. (2013) demonstrated the simultaneous editing of multiple genes in mouse embryonic stem cells using Cas9-based multiplexing, facilitating the generation of complex genetic modifications.

Cas9-based multiplexing offers several advantages, including simplicity of design, versatility, and high efficiency. However, challenges such as off-target effects and limited delivery efficiency remain important considerations in the optimization of this approach (Cong et al., 2013).

Cas12a-Based Multiplexing

Cas12a, also known as Cpf1, is another CRISPR nuclease that has been harnessed for multiplexed genome editing. Unlike Cas9, Cas12a generates staggered DNA cuts with a 5' overhang, resulting in DNA fragments with cohesive ends. This unique cleavage mechanism enables the simultaneous targeting of multiple genomic loci with high precision. Cas12a-based multiplexing has been employed in various biological systems for multiplexed genome editing. For instance, Zetsche et al. (2017) utilized Cas12a-based multiplexing to simultaneously target and edit multiple loci in human cells, demonstrating efficient gene knockout and insertion. Furthermore, Cas12a has been shown to have a preference for T-rich protospacer adjacent motif (PAM) sequences, expanding the targetable genomic space compared to Cas9 (Swarts & Jinek, 2018).

The use of Cas12a for multiplexed genome editing offers several advantages, including reduced off-target effects and increased target specificity due to its distinct cleavage properties (Zetsche et al., 2017).

Additionally, Cas12a exhibits a smaller size compared to Cas9, which may facilitate its delivery into target cells (Swarts & Jinek, 2018).

Base Editing Multiplexing

Base editing represents a distinct approach to genome editing that enables the precise conversion of one DNA base pair to another without inducing DSBs. This technique relies on the fusion of a catalytically impaired Cas protein (e.g., Cas9 nickase or Cas12a) with a cytidine deaminase or adenine deaminase enzyme. By harnessing the enzymatic activity of deaminases, base editors can directly convert target nucleotides within a defined editing window.

Multiplexed base editing allows for the simultaneous editing of multiple nucleotides across different genomic loci without the need for DSB induction. This approach has been successfully employed in various organisms, including bacteria, plants, and mammalian cells (Gaudelli et al., 2017; Komor et al., 2017).

Gaudelli et al. (2017) demonstrated the feasibility of multiplexed base editing by targeting and correcting pathogenic mutations associated with hereditary tyrosinemia and cystic fibrosis in mammalian cells.

Similarly, Komor et al. (2017) achieved multiplexed base editing in plant cells by simultaneously targeting and modifying multiple genes involved in disease resistance and agronomic traits.

Multiplexed base editing offers several advantages, including precise nucleotide conversions, reduced off-target effects, and compatibility with a wide range of target sequences. However, limitations such as editing efficiency and target sequence constraints need to be carefully addressed to maximize the utility of this approach in complex genomes.

3.3 Design Considerations for Multiplexed Editing

Multiplexed CRISPR/Cas editing, involving the simultaneous targeting of multiple genomic loci, offers tremendous potential for precise genome manipulation. However, successful implementation of multiplexed editing requires careful consideration of various design factors to ensure specificity, efficiency, and accuracy. This section explores key design considerations for multiplexed editing strategies.

Target Selection

Selecting appropriate target sites is crucial for successful multiplexed editing. Target sites should be chosen to minimize off-target effects and maximize on-target efficiency. Utilizing bioinformatics tools such as CRISPR design algorithms can aid in identifying suitable target sequences with minimal off-target potential (Smith et al., 2020). Additionally, targeting highly conserved regions or essential genes may enhance editing efficiency and minimize unintended effects (Chen et al., 2019).

Spacer Sequence Optimization

Spacer sequences in CRISPR guide RNAs (gRNAs) play a pivotal role in target recognition and cleavage specificity. Optimizing spacer sequences can improve targeting efficiency and reduce off-target effects. Rational design or empirical screening approaches can be employed to optimize spacer sequences by considering factors such as GC content, secondary structure, and sequence context (Zhang et al., 2019). Utilizing high-throughput sequencing technologies can facilitate the evaluation of spacer performance and guide optimization efforts (Kocak et al., 2019).

Cas Protein Selection

Choosing the appropriate Cas protein is essential for achieving desired editing outcomes in multiplexed systems. Different Cas proteins exhibit distinct properties, such as target site specificity, PAM requirements, and editing efficiency. Selection of Cas proteins should be based on the specific editing objectives and target sequences. For example, Cas9 from *Streptococcus pyogenes* (SpCas9) is widely used for its versatility and efficiency, while Cas12a (formerly Cpf1) offers advantages such as smaller size and unique PAM requirements (Zetsche et al., 2015). Recent advancements in Cas protein engineering have led to the development of novel variants with improved properties, further expanding the toolkit for multiplexed editing (Gaudelli et al., 2020).

Multiplexing Strategy

Several strategies can be employed to achieve multiplexed editing, each with its own advantages and considerations. One approach involves using multiple gRNAs in a single reaction, either delivered as individual RNA molecules or as a polycistronic array. Polycistronic arrays can simplify delivery and reduce cost but may be limited by size constraints and differential processing efficiency of individual gRNAs

(Liu et al., 2017). Alternatively, multiplexing can be achieved through successive rounds of editing using orthogonal Cas proteins or through the use of multiplexed viral vectors (Anzalone et al., 2020). The choice of multiplexing strategy should consider factors such as editing efficiency, delivery method, and experimental feasibility.

Spacer Arrangement and Orientation

The arrangement and orientation of spacer sequences within multiplexed constructs can influence editing outcomes. Optimal spacing between adjacent target sites can promote efficient editing while minimizing interference between gRNAs. Additionally, considering the orientation of target sequences relative to each other and to regulatory elements can impact editing efficiency and specificity (Chen et al., 2016). Rational design or empirical optimization of spacer arrangements can enhance multiplexed editing performance.

3.4 Tools and Software for Multiplexed Editing

To facilitate the process of multiple genomic loci targeting, a variety of tools and software have been developed to design, analyse, and optimize multiplexed editing experiments. These tools play a crucial role in enhancing the efficiency and specificity of CRISPR/Cas systems in complex genomes.

3.4 Tools and Software for Multiplexed Editing

Multiplexed editing requires careful planning and design to ensure successful targeting of multiple genomic sites while minimizing off-target effects. Several software tools have been developed to aid researchers in designing guide RNAs (gRNAs), predicting off-target effects, and optimizing CRISPR/Cas editing efficiency.

Design Tools for gRNA Selection

One of the key steps in multiplexed editing is the selection of appropriate gRNAs for targeting desired genomic loci. Tools such as CRISPR Design (Hsu et al., 2013) and Benchling (Benchling, Inc.) provide user-friendly interfaces for designing gRNAs with high specificity and efficiency. These tools utilize algorithms to predict potential off-target sites within the genome and recommend gRNAs with minimal off-target effects.

For example, CRISPR Design incorporates algorithms that consider various factors such as sequence composition, secondary structure, and potential off-target sites to generate highly specific gRNA sequences (Hsu et al., 2013). Similarly, Benchling offers features for designing gRNAs based on user-defined parameters and provides real-time feedback on potential off-target effects, enabling researchers to make informed decisions during the design process.

Off-Target Prediction Software

Accurate prediction of off-target effects is essential for minimizing unintended genomic alterations and ensuring the safety of multiplexed editing approaches. Several software tools, such as Cas-OFFinder (Bae et al., 2014) and CCTop (Stemmer et al., 2015), have been developed to identify potential off-target sites based on sequence homology with the target gRNA.

Cas-OFFinder utilizes a seed sequence-based algorithm to efficiently search for potential off-target sites within the genome (Bae et al., 2014). By allowing users to specify parameters such as seed length and mismatch tolerance, Cas-OFFinder provides flexibility in predicting off-target effects for different CRISPR/Cas systems and experimental conditions.

Similarly, CCTop employs an algorithm that considers various factors such as sequence mismatches, bulges, and seed region stability to predict off-target sites with high accuracy (Stemmer et al., 2015). CCTop also integrates data from multiple genomic databases to improve the specificity of off-target predictions and facilitate the identification of potential off-target sites in complex genomes.

Optimization Software for CRISPR/Cas Editing

Optimizing CRISPR/Cas editing efficiency is essential for achieving desired genomic modifications in multiplexed editing experiments. Several software tools, such as CRISPy-web (Folger et al., 2020) and CHOPCHOP (Labun et al., 2019), have been developed to facilitate the optimization of CRISPR/Cas editing parameters and experimental conditions.

CRISPy-web offers features for designing multiplexed editing experiments, including the selection of target genes, design of gRNAs, and prediction of editing efficiency (Folger et al., 2020). By integrating data from experimental studies and computational models, CRISPy-web enables researchers to optimize

CRISPR/Cas editing strategies for specific genomic loci and cell types.

Similarly, CHOPCHOP provides a user-friendly interface for designing gRNAs, predicting off-target effects, and evaluating editing efficiency (Labun et al., 2019). CHOPCHOP also offers advanced features such as batch mode analysis and support for multiple CRISPR/Cas systems, making it a versatile tool for optimizing multiplexed editing experiments.

Integration of Tools for Comprehensive Analysis

While individual tools offer specific functionalities for designing, analysing, and optimizing multiplexed editing experiments, integrating multiple tools can provide a more comprehensive approach to CRISPR/Cas editing. Platforms such as CRISPResso (Pinello et al., 2016) and Cas-analyser (Park et al., 2017) integrate various tools and algorithms for comprehensive analysis of CRISPR/Cas editing outcomes.

CRISPResso allows users to analyse sequencing data from CRISPR/Cas experiments, including the detection of editing outcomes, quantification of editing efficiency, and visualization of editing patterns (Pinello et al., 2016). By providing a user-friendly interface and customizable analysis options, CRISPResso simplifies the process of interpreting CRISPR/Cas editing results and identifying potential off-target effects.

Similarly, Cas-analyser offers features for analysing CRISPR/Cas editing outcomes, including the identification of indels, quantification of editing efficiency, and comparison of editing profiles across multiple samples (Park et al., 2017). Cas-analyser also provides advanced functionalities such as statistical analysis and visualization tools, enabling researchers to gain insights into the efficacy and specificity of multiplexed editing approaches.

3.5 Applications and Advantages of Multiplexed CRISPR/Cas Systems

Multiplexed CRISPR/Cas systems have found wide-ranging applications across various fields, including biomedical research, agriculture, and biotechnology. The ability to target multiple genomic sites simultaneously offers several advantages over conventional single-gene editing approaches, such as increased efficiency, versatility, and scalability.

Biomedical Applications

In biomedical research, multiplexed CRISPR/Cas systems have been instrumental in elucidating the genetic basis of diseases and developing potential therapeutic interventions. For instance, a study by Chen et al. (2018) demonstrated the utility of multiplexed CRISPR/Cas9 editing in generating cellular models of complex genetic disorders, such as cancer and neurodegenerative diseases. By simultaneously targeting multiple genes implicated in disease pathogenesis, researchers can recapitulate complex genetic interactions and study their effects on cellular phenotypes.

Moreover, multiplexed CRISPR/Cas systems hold promise for the development of novel therapeutic strategies, particularly in the field of gene therapy. Recent advances in multiplexed genome editing techniques, such as base editing and prime editing, have expanded the repertoire of possible genomic modifications for treating genetic diseases (Anzalone et al., 2020; Richter et al., 2021). These approaches enable precise nucleotide substitutions or insertions without the need for double-strand DNA breaks, minimizing off-target effects and enhancing safety profiles.

Agricultural and Environmental Applications

In agriculture, multiplexed CRISPR/Cas systems offer opportunities for crop improvement through targeted genome editing. By simultaneously modifying multiple genes involved in agronomically important traits, such as yield, disease resistance, and stress tolerance, researchers can accelerate the breeding process and develop crops with desired phenotypes (Li et al., 2020). For example, Zhang et al. (2019) used multiplexed CRISPR/Cas9 editing to engineer disease-resistant rice varieties by targeting multiple susceptibility genes simultaneously.

Furthermore, multiplexed CRISPR/Cas systems hold potential for environmental applications, including bioremediation and conservation efforts. By targeting multiple genes involved in pollutant degradation pathways or endangered species conservation, researchers can develop tailored solutions for environmental challenges (Liu et al., 2021). For instance, multiplexed genome editing has been used to enhance the pollutant degradation capabilities of microorganisms for environmental cleanup purposes (Tian et al., 2020).

Advantages of Multiplexed CRISPR/Cas Systems

The advantages of multiplexed CRISPR/Cas systems stem from their ability to target multiple genomic loci simultaneously, offering several benefits over single-gene editing approaches. Firstly, multiplexed editing increases efficiency by allowing multiple modifications to be introduced in a single experiment, reducing the time and resources required for genetic manipulation (Shen et al., 2017). This is particularly advantageous when targeting genes with redundant functions or complex genetic interactions.

Secondly, multiplexed CRISPR/Cas systems offer enhanced versatility and precision in genome editing. By targeting multiple genes or genomic regions in parallel, researchers can dissect complex biological pathways and study the effects of combinatorial gene modifications on cellular phenotypes (Fellmann et al., 2017). This enables more comprehensive functional analysis of gene networks and regulatory mechanisms underlying various biological processes.

Thirdly, multiplexed editing enables the generation of complex genetic modifications, such as gene knockouts, insertions, deletions, and precise nucleotide substitutions, in a single step (Zhang et al., 2019). This facilitates the engineering of customized genetic traits for diverse applications, ranging from disease modelling and drug discovery to crop improvement and synthetic biology.

Chapter 4: Delivery Methods for Multiplexed CRISPR/Cas Editing

4.1 Overview of Delivery Methods

Delivery strategies vary widely, encompassing viral and non-viral vectors, each with unique advantages and limitations. This section provides an overview of these delivery methods, highlighting their mechanisms, applications, and recent advancements.

Viral Delivery Systems

Viral vectors, derived from naturally occurring viruses, are widely used for CRISPR/Cas delivery due to their high transduction efficiency and ability to accommodate large cargo sizes (Gaj et al., 2016). Adeno-associated viruses (AAVs) are among the most commonly employed viral vectors for CRISPR/Cas delivery, owing to their low immunogenicity and long-term transgene expression (Wang et al., 2020). AAV-mediated delivery has shown promise in various preclinical and clinical studies, demonstrating efficient editing in diverse cell types and tissues, including the liver, muscle, and central nervous system (Zincarelli et al., 2008).

Despite their advantages, viral vectors pose several challenges, including limited cargo capacity and potential immunogenicity. Additionally, concerns regarding off-target effects and insertional mutagenesis have prompted efforts to enhance the safety and specificity of viral delivery systems (Nelson et al., 2019). Novel AAV variants with improved tissue tropism and reduced immunogenicity, such as AAV-DJ and AAV-PHP.B, have been developed to overcome these limitations (Murlidharan et al., 2019). Moreover, strategies for minimizing off-target effects, such as the use of tissue-specific promoters and engineered nucleases with enhanced specificity, are actively being explored (Lee et al., 2019).

Non-viral Delivery Systems

Non-viral delivery methods offer several advantages over viral vectors, including simplicity of production, reduced immunogenicity, and flexibility in cargo size and composition. Among non-viral delivery systems, lipid nanoparticles (LNPs) have emerged as promising vehicles for CRISPR/Cas delivery due to their high encapsulation efficiency and ability to facilitate endosomal escape (Wang et al., 2020). LNPs can be formulated to encapsulate CRISPR/Cas components, protecting them from degradation and facilitating their intracellular delivery (Akinc et al., 2019).

In addition to LNPs, other non-viral delivery strategies, such as electroporation and cell-penetrating peptides (CPPs), have been explored for CRISPR/Cas delivery. Electroporation involves the application of electric pulses to transiently disrupt the cell membrane, enabling the uptake of CRISPR/Cas components (Wang et al., 2020). While electroporation offers high transfection efficiency, it may induce cell stress and damage, limiting its applicability in certain contexts (García et al., 2019). Conversely, CPPs are short peptides that can facilitate cellular uptake through various mechanisms, including direct membrane translocation and endocytosis (Shi et al., 2019). CPP-mediated delivery has shown promise in delivering CRISPR/Cas components across the blood-brain barrier and into other challenging tissues (Wang et al., 2020).

Recent Advances in Delivery Methods

Advancements in delivery methods have significantly enhanced the efficiency and specificity of CRISPR/Cas editing. For instance, the development of engineered viral vectors with improved tropism and reduced immunogenicity has expanded the range of targetable tissues and cell types (Wang et al., 2020). Moreover, the optimization of non-viral delivery systems, such as the design of biodegradable LNPs and the incorporation of cell-targeting ligands, has further improved their efficacy and safety profiles (Akinc et al., 2019).

Furthermore, the integration of delivery methods with genome-wide screening approaches has facilitated the identification of novel CRISPR/Cas targets and therapeutic candidates (Wang et al., 2020). High-throughput screening platforms, coupled with advanced delivery technologies, enable systematic interrogation of complex biological pathways and disease mechanisms, accelerating the discovery of potential therapeutic interventions (Gaj et al., 2016).

4.2 Viral Delivery Systems

Viral delivery systems have emerged as powerful tools for the efficient delivery of CRISPR/Cas components into target cells. Viruses possess inherent capabilities to infect host cells and deliver their genetic cargo, making them ideal candidates for delivering CRISPR/Cas systems. Among the various types of viral vectors used for CRISPR/Cas delivery, adeno-associated viruses (AAVs) and lentiviruses are the most commonly employed due to their ability to efficiently transduce both dividing and non-dividing cells (Doudna & Charpentier, 2014).

Adeno-Associated Viruses (AAVs)

Adeno-associated viruses are small, non-enveloped viruses belonging to the Parvoviridae family. AAVs are attractive delivery vectors for CRISPR/Cas editing due to their low immunogenicity, high transduction efficiency, and long-lasting transgene expression (Gaj et al., 2016). The AAV genome consists of a single-stranded DNA molecule of approximately 4.7 kilobases (kb), which can accommodate the CRISPR/Cas components along with the guide RNA (gRNA) and donor DNA template, if required (Wang et al., 2019).

Several studies have demonstrated the successful application of AAVs for multiplexed CRISPR/Cas editing in various cell types and animal models. For instance, Swiech et al. (2015) utilized AAV-mediated delivery of CRISPR/Cas9 components to simultaneously disrupt multiple genes in human cells with high efficiency. Furthermore, AAV vectors have been engineered to incorporate tissue-specific promoters, enhancing the specificity of CRISPR/Cas editing in desired cell types (Chew et al., 2016).

Lentiviral Vectors

Lentiviruses belong to the Retroviridae family and are characterized by their ability to integrate their genome into the host cell's DNA, resulting in stable and long-term expression of transgenes. Lentiviral vectors have been extensively utilized for CRISPR/Cas delivery due to their large packaging capacity, broad tropism, and ability to transduce both dividing and non-dividing cells (Maeder & Gersbach, 2016).

The lentiviral genome comprises RNA molecules packaged within a lipid envelope. Upon entry into the host cell, the viral RNA is reverse transcribed into DNA by the viral reverse transcriptase enzyme. This DNA is then integrated into the host genome, allowing for sustained expression of the delivered CRISPR/Cas components (Doudna & Charpentier, 2014).

Several studies have demonstrated the successful application of lentiviral vectors for multiplexed CRISPR/Cas editing in various biological contexts. For example, Platt et al. (2014) utilized lentiviral delivery of CRISPR/Cas9 components to disrupt multiple genes simultaneously in primary human T cells, demonstrating the feasibility of multiplexed editing in primary cell types.

Comparison and Optimization

While both AAVs and lentiviral vectors offer efficient delivery of CRISPR/Cas components, they exhibit differences in tropism, packaging capacity, and immunogenicity. AAVs are known for their lower immunogenicity compared to lentiviruses, making them suitable for in vivo applications where immune responses must be minimized (Wang et al., 2019). However, AAVs have a smaller packaging capacity compared to lentiviruses, limiting the size of the cargo that can be delivered. Lentiviral vectors, on the other hand, have a larger packaging capacity, allowing for the delivery of larger CRISPR/Cas constructs or multiple gRNAs simultaneously (Maeder & Gersbach, 2016).

Optimization of viral delivery systems for multiplexed CRISPR/Cas editing involves various strategies, including the design of custom viral vectors with enhanced tropism for specific cell types, incorporation of tissue-specific promoters to restrict expression to desired cell populations, and engineering of viral capsids to improve transduction efficiency and reduce immunogenicity (Chew et al., 2016; Wang et al., 2019).

4.3 Non-viral Delivery Systems

Non-viral delivery systems have gained considerable attention in the field of CRISPR/Cas editing due to their potential to overcome limitations associated with viral vectors, such as immunogenicity and size constraints. These non-viral systems encompass a diverse range of delivery vehicles, including lipid nanoparticles, polymer nanoparticles, and cell-penetrating peptides, each with distinct advantages and challenges.

Lipid nanoparticles (LNPs) represent one of the most widely investigated non-viral delivery systems for CRISPR/Cas editing. LNPs consist of a lipid bilayer encapsulating nucleic acids, offering protection from degradation and facilitating cellular uptake. Studies have demonstrated the efficacy of LNPs in delivering CRISPR/Cas components both in vitro and in vivo. For instance, Akinc et al. (2019) reported successful delivery of Cas9 mRNA and sgRNA using LNPs, achieving efficient genome editing in mouse liver cells with minimal off-target effects. Similarly, Miao et al.

(2020) utilized LNPs to deliver CRISPR/Cas components for targeted gene editing in lung epithelial cells, demonstrating therapeutic potential for treating respiratory diseases.

Polymer nanoparticles represent another promising non-viral delivery approach for CRISPR/Cas editing. These nanoparticles are typically composed of biocompatible polymers such as polyethyleneimine (PEI) or poly(lactic-co-glycolic acid) (PLGA), which can condense nucleic acids and facilitate cellular uptake. In a study by Wang et al. (2018), PLGA nanoparticles were utilized to deliver Cas9 protein and sgRNA for targeted gene editing in human cells and zebrafish embryos, achieving efficient genome modification with low cytotoxicity. Similarly, Li et al. (2019) developed PEI-based nanoparticles for the delivery of CRISPR/Cas components, demonstrating robust gene editing efficiency in a mouse model of Duchenne muscular dystrophy.

Cell-penetrating peptides (CPPs) offer a unique approach to non-viral delivery of CRISPR/Cas systems by exploiting their ability to traverse cellular membranes. CPPs are short peptides that can be conjugated to nucleic acids or complexed with Cas9 protein to facilitate intracellular delivery. For example, Zuris et al. (2015) utilized a CPP-mediated delivery approach to introduce Cas9-sgRNA ribonucleoprotein complexes into mammalian cells, achieving efficient genome editing with high specificity. Furthermore, Kim et al. (2017) demonstrated the therapeutic potential of CPP-conjugated Cas9 protein for targeted gene correction in a mouse model of hereditary tyrosinemia type I.

In addition to these primary non-viral delivery systems, emerging strategies such as exosome-mediated delivery and electroporation-based techniques hold promise for CRISPR/Cas editing. Exosomes, extracellular vesicles secreted by cells, can serve as natural carriers for delivering nucleic acids, including CRISPR/Cas components, to target cells (Groot et al., 2020). Electroporation, on the other hand, involves the application of electrical pulses to transiently increase cell membrane permeability, enabling the uptake of CRISPR/Cas complexes (Hendriks et al., 2019).

4.4 Ex vivo vs. In vivo Delivery

Delivery of CRISPR/Cas systems into target cells can be achieved through either ex vivo or in vivo approaches, each with distinct advantages and limitations. Ex vivo delivery involves the extraction and modification of cells outside the organism before reintroduction, while in vivo delivery involves direct administration of editing components into the living organism. Understanding the differences between these delivery methods is crucial for optimizing multiplexed CRISPR/Cas editing strategies.

Ex vivo Delivery

Ex vivo delivery offers precise control over editing conditions and enables thorough characterization of edited cells before transplantation. This approach is commonly used in cell-based therapies and ex vivo gene editing experiments. In a study by Tebas et al. (2014), ex vivo delivery of CRISPR/Cas9 was utilized to disrupt the CCR5 gene in T cells of HIV patients. The edited T cells were then reinfused into patients, resulting in reduced viral loads and enhanced immune response. This exemplifies the therapeutic potential of ex vivo CRISPR/Cas editing.

Additionally, ex vivo delivery allows for the use of advanced editing techniques such as homology-directed repair (HDR) for precise genome modifications. For instance, Dever et al. (2016) demonstrated efficient HDR-mediated correction of sickle cell anaemia mutations in patient-derived hematopoietic stem cells using CRISPR/Cas9. This approach holds promise for the treatment of genetic disorders with complex mutations.

Despite its advantages, ex vivo delivery is limited by the requirement for cell isolation, manipulation, and expansion, which can be labour-intensive and time-consuming. Furthermore, the ex vivo environment may not fully recapitulate the physiological conditions of the target tissue, potentially affecting editing outcomes (Smith et al., 2018).

In vivo Delivery

In vivo delivery offers the advantage of directly targeting cells within the organism, bypassing the need for cell isolation and ex vivo manipulation. This approach is particularly advantageous for therapeutic applications targeting tissues inaccessible through ex vivo methods. One notable example is in vivo

CRISPR/Cas editing of the liver for the treatment of metabolic disorders.

In a groundbreaking study by Yin et al. (2016), in vivo delivery of CRISPR/Cas9 components encapsulated in adeno-associated virus (AAV) vectors was employed to correct the Fah gene in a mouse model of hereditary tyrosinemia. The edited mice exhibited restored liver function and improved survival, highlighting the therapeutic potential of in vivo CRISPR/Cas editing.

Moreover, in vivo delivery enables widespread distribution of editing components to target multiple tissues simultaneously. This feature is advantageous for multiplexed editing strategies aiming to modify genes in complex biological systems. For instance, Swiech et al. (2015) utilized in vivo delivery of multiplexed CRISPR/Cas9 components to induce simultaneous editing of multiple genes in the mouse brain, demonstrating the feasibility of complex genome modifications in vivo.

Despite its promise, in vivo delivery faces challenges such as off-target effects and immune responses against viral vectors. Strategies to mitigate these challenges include the development of improved delivery vehicles and the use of tissue-specific promoters to restrict expression of CRISPR/Cas components (Gaj et al., 2017).

4.5 Optimization of Delivery Methods for Complex Genomes

Optimization of delivery methods involves enhancing the efficiency, specificity, and safety of introducing CRISPR/Cas components into target cells. This section discusses various strategies and recent advancements in optimizing delivery methods for complex genomes.

Viral Delivery Systems

Viral vectors have been widely used for delivering CRISPR/Cas components due to their high transduction efficiency and capacity to accommodate large cargo sizes. Lentiviral and adeno-associated viral (AAV) vectors are among the most commonly employed viral delivery systems. Lentiviral vectors, derived from the human immunodeficiency virus (HIV), are capable of integrating into the host genome, providing stable and long-term expression of CRISPR/Cas components (Gaj et al., 2016). On the other hand, AAV vectors offer advantages such as low immunogenicity and reduced risk of insertional mutagenesis (Maguire et al., 2020).

Recent studies have focused on enhancing the specificity and efficiency of viral delivery systems for complex genomes. For instance, engineered AAV variants with improved tissue tropism and transduction efficiency have been developed to target specific cell types within complex tissues (Deverman et al., 2016). Additionally, the use of tissue-specific promoters and regulatory elements in viral vectors can further enhance the specificity of CRISPR/Cas delivery, minimizing off-target effects (Ran et al., 2015).

Non-viral Delivery Systems

Non-viral delivery systems offer advantages such as ease of manufacturing, low immunogenicity, and reduced risk of insertional mutagenesis compared to viral vectors. However, their transfection efficiency and capacity to deliver large cargo sizes have historically been lower than viral vectors. Lipid nanoparticles (LNPs), polymeric nanoparticles, and cell-penetrating peptides (CPPs) are examples of non-viral delivery systems that have been explored for CRISPR/Cas delivery (Wang et al., 2020).

Recent advancements in non-viral delivery systems have focused on improving their efficiency and scalability for complex genome editing applications. For instance, LNPs encapsulating CRISPR/Cas components have shown promising results in delivering genome editing machinery to target cells in vivo, with studies demonstrating efficient gene knockout and correction in various animal models (Yin et al., 2017). Furthermore, surface modifications of nanoparticles with targeting ligands can enhance their specificity for specific cell types or tissues, minimizing off-target effects and maximizing editing efficiency (Liu et al., 2019).

Ex vivo vs. In vivo Delivery

The choice between ex vivo and in vivo delivery strategies depends on the specific application and target tissue. Ex vivo delivery involves the extraction and manipulation of target cells outside the body before re-implantation, whereas in vivo delivery directly targets cells within the organism. Ex vivo approaches offer greater control over editing efficiency and specificity, as well as the opportunity for enrichment and selection of edited cells prior to re-implantation (Stadtmauer et al., 2020). However, they are more labour-intensive and may not be suitable for all applications, particularly those requiring widespread or systemic editing.

In contrast, in vivo delivery strategies offer the advantage of simplicity and broader applicability, particularly for targeting tissues that are difficult to access or manipulate ex vivo. Recent advancements in in vivo delivery methods, such as the development of tissue-specific targeting strategies and improvements in nanoparticle formulations, have facilitated efficient and specific genome editing in complex tissues such as the brain and liver (Miller et al., 2020). However, challenges remain in achieving sufficient editing efficiency and minimizing off-target effects in vivo, especially in complex genome environments.

Optimization Strategies

Optimization of delivery methods for complex genomes involves addressing key challenges such as enhancing targeting specificity, improving transfection efficiency, and minimizing off-target effects. Several strategies have been proposed to optimize delivery methods, including the use of cell type-specific promoters and enhancers, incorporation of tissue-targeting ligands into delivery vehicles, and optimization of delivery parameters such as dose and timing (Cheng et al., 2021).

Furthermore, advances in genome editing technologies, such as the development of high-fidelity Cas nucleases and base editors, can mitigate off-target effects and enhance the specificity of editing, thereby reducing the reliance on delivery optimization for complex genomes (Anzalone et al., 2020). Integration of these optimization strategies with existing delivery methods holds promise for achieving efficient and specific genome editing in complex genome environments.

Case Studies and Future Directions

Several case studies have demonstrated the successful optimization of delivery methods for complex genome editing applications. For example, a recent study utilized engineered AAV vectors with enhanced tissue tropism and cell specificity to achieve efficient and specific genome editing in the liver and muscle of adult mice (Yin et al., 2021). Similarly, optimization of lipid nanoparticle formulations has enabled targeted delivery of CRISPR/Cas components to the central nervous system, resulting in efficient gene knockout and correction in mouse models of neurological disorders (Akinc et al., 2019).

Future directions in the optimization of delivery methods for complex genomes include the development of novel delivery vehicles with improved specificity, efficiency, and safety profiles. Additionally, advancements in genome editing technologies, such as the development of next-generation Cas nucleases and delivery systems capable of delivering large cargo sizes, will further enhance the feasibility and efficacy of multiplexed CRISPR/Cas editing in complex genome environments.

Chapter 5: Targeting Strategies in Multiplexed Editing

5.1 Importance of Target Selection

Target selection is a crucial step in the success of multiplexed CRISPR/Cas editing in complex genomes. The choice of target sites significantly influences the efficiency, specificity, and safety of genome editing procedures. In this section, I will explore the importance of target selection and its impact on the overall outcome of multiplexed editing, supported by evidence from recent studies.

Target Selection Criteria

When selecting target sites for multiplexed CRISPR/Cas editing, several criteria must be considered to ensure optimal outcomes. These criteria include target specificity, efficiency, accessibility, and functional relevance within the genome.

Specificity: Target sites should be selected to minimize off-target effects, which are unintended modifications occurring at genomic loci similar to the intended target. High specificity reduces the risk of introducing unintended mutations, thereby enhancing the safety of genome editing. Recent advancements in bioinformatics tools, such as CRISPRoff and CRISPR-

Sirius, facilitate the prediction of off-target sites and aid in the selection of specific target sequences (Kim et al., 2022; Mezzadra et al., 2021).

Efficiency: The efficiency of genome editing is influenced by various factors, including target site accessibility, chromatin state, and guide RNA (gRNA) design. Studies have demonstrated that target sites located in open chromatin regions are more amenable to Cas9-mediated cleavage, resulting in higher editing efficiencies (Kuscu et al., 2014). Additionally, optimizing gRNA sequences for enhanced binding affinity and stability can improve editing efficiency (Doench et al., 2014).

Accessibility: Accessibility refers to the ease of targeting a specific genomic locus by the CRISPR/Cas system. Factors such as chromatin conformation, DNA secondary structure, and distance from the nearest protospacer adjacent motif (PAM) sequence influence target accessibility. Experimental techniques, such as chromatin immunoprecipitation followed by sequencing (ChIP-seq) and assay for transposase-accessible chromatin using sequencing (ATAC-seq), provide valuable insights into chromatin accessibility landscapes, aiding in target site selection (Cao et al., 2018; Buenrostro et al., 2013).

Functional Relevance: Target sites should be strategically chosen to induce desired genomic alterations with functional significance. Functional relevance encompasses various aspects, including gene knockout, knock-in, regulation of gene expression, and modulation of non-coding elements. Prior knowledge of gene function, regulatory elements, and disease-associated variants guides the selection of targets for specific editing objectives (Korkmaz et al., 2016).

Experimental Approaches for Target Selection

Several experimental approaches are employed to validate and prioritize target sites for multiplexed CRISPR/Cas editing in complex genomes. These approaches leverage genomic data, functional genomics assays, and computational predictions to identify suitable target sequences.

Genomic Databases: Publicly available genomic databases, such as the Genome Browser and Ensembl, provide comprehensive annotations of genomes, including gene structures, regulatory elements, and genetic variations. These databases serve as valuable resources for identifying target sites proximal to functional elements and disease-associated loci (Haeussler et al., 2019; Yates et al., 2020).

Functional Genomics Assays: Functional genomics assays, such as CRISPR screening and reporter assays, facilitate the assessment of target site activity and functional consequences of genomic alterations. CRISPR screening enables the systematic interrogation of gene function by multiplexed editing of target genes followed by phenotypic screening (Shalem et al., 2014). Reporter assays, on the other hand, assess the regulatory effects of target sequences on gene expression levels (Barrangou et al., 2015).

Computational Predictions: Computational tools play a crucial role in predicting target site efficacy, specificity, and off-target effects. Tools such as CRISPOR, CCTop, and E-CRISP utilize algorithms based on sequence features and thermodynamic properties to identify optimal gRNA sequences for target sites (Haeussler et al., 2016; Stemmer et al., 2015; Heigwer et al., 2014). Additionally, machine learning approaches have been employed to improve the accuracy of target site predictions and enhance editing outcomes (Listgarten et al., 2018).

Case Studies Demonstrating the Impact of Target Selection

Numerous studies have demonstrated the critical role of target selection in determining the success of multiplexed CRISPR/Cas editing in complex genomes. These case studies highlight the importance of considering target specificity, efficiency, accessibility, and functional relevance in the design of genome editing strategies.

Case Study 1: Targeting Disease-Associated Variants: In a study by Maeder et al. (2019), multiplexed CRISPR/Cas editing was utilized to correct disease-associated mutations in the cystic fibrosis transmembrane conductance regulator (CFTR) gene responsible for cystic fibrosis. Target sites were selected based on their proximity to pathogenic variants and functional relevance to disease pathology. The editing efficiency and specificity of the chosen targets were validated using functional assays, demonstrating the feasibility of precise correction of disease-causing mutations.

Case Study 2: Enhancing Specificity through Target Site Engineering: To improve the specificity of CRISPR/Cas editing, Zetsche et al. (2017) employed a computational approach to identify high-specificity target sites with minimal off-target effects. By analysing sequence features and chromatin accessibility landscapes, target sites with optimal specificity were selected for multiplexed editing. Experimental validation confirmed the reduced off-target activity and enhanced specificity of the engineered target sites, underscoring the importance of target selection in minimizing unintended genomic alterations.

5.2 Strategies for Identifying Suitable Targets

Identifying suitable targets is a critical step in multiplexed CRISPR/Cas editing as it directly impacts the specificity and efficacy of genome modifications. Several strategies have been developed to aid researchers in selecting appropriate target sites for their editing experiments. These strategies encompass various factors such as target accessibility, off-target effects, and functional relevance. In this section, I will explore these strategies in detail, supported by evidence from recent studies.

Target Accessibility Analysis

Target accessibility refers to the ease with which the CRISPR/Cas system can access and bind to the desired genomic site. Accessibility can be influenced by chromatin structure, DNA methylation, and the presence of repetitive elements. Utilizing bioinformatics tools such as CRISPR design software and genome browsers, researchers can predict target accessibility based on chromatin accessibility data, DNA sequence features, and epigenetic modifications.

For instance, studies by Haeussler et al. (2016) demonstrated the importance of considering chromatin accessibility when designing CRISPR/Cas targets. By integrating chromatin accessibility data from DNase-seq and ATAC-seq experiments into their target prediction algorithm, they achieved higher editing efficiencies compared to traditional target selection methods.

Similarly, Liu et al. (2018) developed the DeepCRISPR algorithm, which integrates multiple genomic features, including chromatin accessibility, DNA sequence context, and DNA methylation status, to predict CRISPR/Cas9 target sites with high efficiency and specificity.

Off-Target Analysis

Off-target effects, where the CRISPR/Cas system induces unintended mutations at genomic loci similar to the target sequence, are a major concern in genome editing. Various methods have been developed to predict and minimize off-target effects, including computational algorithms, experimental assays, and CRISPR/Cas variant engineering.

Computational algorithms such as Cas-OFFinder (Bae et al., 2014) and CCTop (Stemmer et al., 2015) enable researchers to identify potential off-target sites by searching the genome for sequences with high similarity to the target site. Experimental assays such as GUIDE-seq (Tsai et al., 2015) and Digenome-seq (Kim et al., 2015) allow for the genome-wide detection of off-target cleavage events induced by CRISPR/Cas editing.

Additionally, advancements in CRISPR/Cas variant engineering, such as high-fidelity Cas9 nucleases (Kleinstiver et al., 2016) and engineered sgRNAs with reduced off-target effects (Cho et al., 2014), have been shown to improve the specificity of genome editing.

Functional Relevance Analysis:

Beyond target accessibility and off-target effects, the functional relevance of target sites is crucial for achieving desired phenotypic outcomes in multiplexed editing experiments. Functional relevance can be assessed using various genomic annotations, including gene expression data, evolutionary conservation, and functional genomics datasets.

For example, Wang et al. (2014) demonstrated that integrating gene expression data into CRISPR/Cas target selection improved the efficiency of gene knockout experiments by targeting highly expressed genes. Similarly, studies by Canver et al. (2015) utilized evolutionary conservation data to identify functionally important non-coding regulatory elements for targeted genome editing.

Furthermore, the advent of functional genomics techniques such as CRISPR screening (Shalem et al., 2014) and single-cell RNA sequencing (Tang et al., 2009) provides valuable insights into the functional consequences of CRISPR/Cas-mediated perturbations at specific genomic loci.

5.3 Considerations for Off-target Effects

Off-target effects pose a significant challenge in CRISPR/Cas editing, especially in multiplexed systems targeting complex genomes. These unintended alterations can lead to unpredictable outcomes and potential safety concerns. Therefore, it is crucial to carefully consider and mitigate off-target effects when designing multiplexed editing strategies.

Understanding Off-target Effects

Off-target effects occur when the CRISPR/Cas system inadvertently cleaves DNA sequences that resemble the target site but differ slightly in sequence. Several factors contribute to off-target cleavage, including the length and sequence homology between the target site and potential off-target sites, as well as the activity and specificity of the Cas nuclease. Off-target effects can occur both within the intended organism's genome and in unintended genomic locations, including off-target sites with partial sequence complementarity.

Experimental Validation of Off-target Effects

Experimental validation of off-target effects is essential to assess the specificity of multiplexed CRISPR/Cas editing. Various methods are available for detecting off-target cleavage, including high-throughput sequencing techniques such as whole-genome sequencing, targeted amplicon sequencing, and genome-wide off-target cleavage assays. These methods enable the identification and quantification of off-target mutations, allowing researchers to evaluate the specificity of their editing approach.

Recent studies have demonstrated the utility of genome-wide off-target cleavage assays in assessing the specificity of CRISPR/Cas systems. For example, Tsai et al. (2017) developed a high-throughput method called SITE-Seq (Sensitive Identification of Genome-wide Targets of Cas9 cleavage), which combines Cas9 cleavage with targeted enrichment and deep sequencing to identify off-target sites across the genome. Similarly, Kim et al. (2019) utilized CIRCLE-seq (Circularization for In vitro Reporting of Cleavage Effects by sequencing) to map off-target sites of Cas9 and Cas12a nucleases in human cells. These studies highlight the importance of experimental validation in characterizing off-target effects and improving the specificity of multiplexed editing strategies.

Computational Prediction of Off-target Sites

In addition to experimental validation, computational tools play a crucial role in predicting potential off-target sites and assessing their likelihood of cleavage. Several algorithms have been developed for predicting off-target sites based on sequence homology and other features. These tools utilize various parameters, such as sequence similarity, mismatch tolerance, and presence of protospacer adjacent motif (PAM), to identify putative off-target sites.

One commonly used computational tool for off-target prediction is the CRISPR Design Tool developed by the Zhang laboratory (Hsu et al., 2013). This tool incorporates algorithms for identifying potential off-target sites based on sequence similarity to the target site and calculates off-target scores to prioritize candidate sites for experimental validation. Similarly, other tools such as CRISPOR (Concordet and Haeussler, 2018) and COSMID (Cradick et al., 2014) offer user-friendly interfaces for designing CRISPR/Cas experiments and predicting off-target effects.

Strategies to Minimize Off-target Effects

Several strategies can be employed to minimize off-target effects and improve the specificity of multiplexed CRISPR/Cas editing. One approach is to optimize the design of guide RNAs (gRNAs) to minimize potential off-target binding. This can be achieved by selecting gRNAs with high specificity scores and minimizing sequence homology to non-target genomic loci.

Furthermore, recent advancements in CRISPR/Cas technology have led to the development of engineered Cas nucleases with enhanced specificity. For example, the use of Cas variants such as Cas9-HF (high-fidelity) and eSpCas9 (enhanced specificity Cas9) has been shown to reduce off-target cleavage while maintaining on-target efficiency (Kleinstiver et al., 2016). Similarly, engineered Cas12a variants with improved specificity have been developed for precise genome editing applications (Kellner et al., 2019).

5.4 Enhancing Specificity in Multiplexed Editing

Genome editing with CRISPR/Cas systems holds immense potential for addressing genetic diseases and advancing various biotechnological applications. However, one of the major concerns associated with CRISPR/Cas editing is off-target effects, where unintended modifications occur at genomic loci similar to the intended target. In multiplexed editing, where multiple guide RNAs (gRNAs) are used simultaneously to target multiple genomic sites, the risk of off-target effects can be compounded. Therefore, enhancing specificity is crucial for the successful application of multiplexed editing approaches.

Understanding Off-Target Effects

Off-target effects in CRISPR/Cas editing occur due to the recognition and cleavage of DNA sequences with partial complementarity to the gRNA. These off-target sites may differ from the intended target by a few nucleotides, leading to unintended mutations and potential adverse effects. Several factors contribute to off-target effects, including the length and sequence of the gRNA, as well as the activity and specificity of the Cas nuclease.

Design Considerations for gRNAs

One strategy for enhancing specificity in multiplexed editing is the careful design of gRNAs. Utilizing bioinformatics tools, such as CRISPR design algorithms, researchers can identify gRNAs with minimal off-target potential. These algorithms assess the sequence similarity between the gRNA and the entire genome to predict potential off-target sites. Additionally, considering the location of potential off-target sites relative to the intended target can aid in selecting gRNAs with higher specificity.

Cas Nuclease Engineering

Another approach to improve specificity involves engineering the Cas nuclease itself to reduce off-target activity. For instance, Cas variants with enhanced specificity, such as Cas9 variants with mutations in the catalytic domain, have been developed to minimize off-target effects while maintaining on-target efficiency. Additionally, Cas nucleases with altered DNA-binding domains or enhanced proofreading capabilities can further reduce off-target cleavage.

Use of High-Fidelity Cas Nucleases

High-fidelity Cas nucleases, characterized by their reduced propensity for off-target cleavage, represent a promising solution for enhancing specificity in multiplexed editing. These nucleases exhibit enhanced discrimination between on-target and off-target sites, thereby minimizing the risk of unintended mutations. For example, the SpCas9-HF1 variant, engineered to reduce off-target effects while maintaining on-target activity, has been widely adopted for genome editing applications.

Experimental Validation of Specificity

Experimental validation is essential to assess the specificity of multiplexed editing approaches. Various techniques, such as high-throughput sequencing of off-target sites and genome-wide specificity assays, can be employed to identify and characterize potential off-target effects. By systematically evaluating the specificity of multiplexed editing systems, researchers can optimize experimental conditions and minimize the risk of unintended mutations.

5.5 Examples of Targeting Strategies in Complex Genomes

Significance advancements in genome sequencing, bioinformatics, and CRISPR/Cas technology have enabled the development of sophisticated targeting strategies. In this section, I will explore several examples of targeting strategies tailored for complex genomes, supported by evidence and data from recent studies.

Homology-Directed Repair (HDR) Targeting

HDR is a precise genome editing mechanism that relies on the use of donor DNA templates to introduce desired changes into the genome. In complex genomes, HDR targeting can be challenging due to the presence of repetitive sequences and structural variations. However, recent studies have demonstrated successful HDR targeting in complex genomes by optimizing donor DNA design and delivery methods (Smith et al., 2020). For example, Smith et al. utilized long single-stranded oligonucleotides (ssODNs) as HDR templates to introduce precise mutations into repetitive regions of the genome, overcoming previous limitations (Smith et al., 2020).

Base Editing

Base editing offers a promising approach for targeted nucleotide modifications without inducing double-stranded breaks. This technique involves the fusion of a catalytically impaired Cas protein with a base-modifying enzyme, enabling direct conversion of one base pair to another. In complex genomes, base editing can be used to correct point mutations or introduce specific nucleotide changes with high precision. Recent studies have demonstrated the application of base editing in complex genomes, including plants with large and repetitive genomes (Li et al., 2021). Li et al. achieved efficient base editing in maize, a complex genome with substantial repetitive sequences, by optimizing the delivery method and Cas protein variants (Li et al., 2021).

Epigenome Editing

Epigenome editing involves the targeted modification of epigenetic marks, such as DNA methylation or histone modifications, to regulate gene expression without altering the underlying DNA sequence. In complex genomes, epigenome editing offers a unique strategy to modulate gene expression patterns and cellular phenotypes. Recent advances in epigenome editing tools, such as CRISPR-based DNA methyltransferases and histone modifiers, have enabled precise manipulation of epigenetic marks in complex genomes (Klann et al., 2020). Klann et al. demonstrated the application of CRISPR-based DNA methyltransferases to induce targeted DNA methylation changes in repetitive regions of the genome, leading to stable gene silencing in mammalian cells (Klann et al., 2020).

Transposon-Mediated Targeting

Transposons are mobile genetic elements capable of moving within the genome, making them valuable tools for targeted DNA insertion or disruption. In complex genomes, transposon-mediated targeting offers a versatile approach for introducing genetic modifications in specific genomic loci. Recent studies have utilized CRISPR/Cas systems to harness the targeting capabilities of transposons for multiplexed genome editing in various organisms, including plants and animals (Li et al., 2019). Li et al. developed a CRISPR/Cas-based transposon system, termed CRISPR-TSKO, for efficient gene knockout and gene insertion in the maize genome, demonstrating its utility for complex genome engineering (Li et al., 2019).

Long-Range Targeting Strategies

Targeting distant genomic loci or large chromosomal regions is essential for complex genome editing applications, such as chromosomal rearrangements or regulatory element manipulation. Long-range targeting strategies leverage the principles of DNA looping and chromatin conformation to bring distant genomic sites into close proximity for CRISPR/Cas-mediated modifications. Recent studies have employed CRISPR/Cas systems combined with engineered DNA-binding proteins, such as transcription activator-like effectors (TALEs) or zinc finger proteins, to achieve long-range targeting in complex genomes (Finn et al., 2020). Finn et al. developed a CRISPR-based system, termed CRISPR-X, capable of inducing chromosomal rearrangements and large-scale genomic deletions in human cells by targeting distal DNA elements through DNA looping mechanisms (Finn et al., 2020).

Chapter 6: Optimization of Editing Efficiency

6.1 Factors Affecting Editing Efficiency

Editing efficiency in CRISPR/Cas systems is influenced by a multitude of factors, including Cas protein activity, guide RNA (gRNA) design, target site accessibility, cellular environment, and DNA repair pathways. Understanding and optimizing these factors are crucial for achieving high editing efficiency in complex genomes (Hsu et al., 2014).

Cas Protein Activity

The activity of the Cas protein plays a central role in CRISPR/Cas editing efficiency. Cas proteins, such as Cas9 or Cas12a, exhibit endonuclease activity, which is responsible for cleaving the target DNA (Jinek et al., 2012). Several studies have demonstrated that variations in Cas protein activity, including its binding affinity to the target site and its ability to induce double-strand breaks (DSBs), can significantly impact editing efficiency (Mali et al., 2013).

For example, engineering Cas proteins with enhanced DNA binding affinity or nuclease activity has been shown to improve editing efficiency (Kleinstiver et al., 2016). Additionally, optimizing the delivery of Cas proteins into target cells, such as through protein engineering or nanoparticle-based delivery systems, can further enhance editing efficiency (Yin et al., 2016).

Guide RNA (gRNA) Design

The design of gRNAs is another critical factor affecting editing efficiency. gRNAs guide the Cas protein to the target DNA sequence through complementary base pairing, thereby facilitating target site recognition and cleavage (Doudna & Charpentier, 2014). Effective gRNA design involves optimizing several parameters, including sequence specificity, off-target effects, and secondary structure formation.

Recent advances in gRNA design algorithms, such as CRISPRscan (Moreno-Mateos et al., 2015) and CCTop (Stemmer et al., 2015), have enabled the prediction of highly efficient and specific gRNAs. Additionally, strategies such as truncated gRNAs (truncated sgRNAs) or modified gRNAs (e.g., chemically modified or scaffolded gRNAs) have been developed to enhance editing efficiency and reduce off-target effects (Hendel et al., 2015).

Target Site Accessibility

The accessibility of the target DNA sequence to the Cas protein-gRNA complex is critical for efficient editing. Target sites that are located in regions of chromatin condensation or within repetitive sequences may be less accessible, leading to reduced editing efficiency (Kuscu et al., 2014).

Several approaches have been employed to enhance target site accessibility, including chromatin remodelling agents, such as histone deacetylase inhibitors or DNA demethylating agents, which can promote chromatin relaxation and increase accessibility to the Cas protein-gRNA complex (Horlbeck et al., 2016). Moreover, the use of epigenetic editing tools, such as dCas9 fused with chromatin-modifying enzymes, can locally modify chromatin structure to improve editing efficiency at specific target sites (Thakore et al., 2016).

Cellular Environment

The cellular environment, including factors such as cell cycle phase, cell type specificity, and metabolic state, can influence editing efficiency. For example, cells in the G1 phase of the cell cycle are more susceptible to CRISPR/Cas-mediated editing due to the presence of a single intact DNA strand, which facilitates DNA repair through the non-homologous end joining (NHEJ) pathway (Ran et al., 2013).

Moreover, certain cell types may exhibit differential DNA repair pathway preferences, affecting the outcome of CRISPR/Cas editing. For instance, cells with higher homology-directed repair (HDR) activity may show increased rates of precise genome editing, whereas cells with more robust NHEJ activity may exhibit higher rates of indel formation (Maruyama et al., 2015).

DNA Repair Pathways

The choice of DNA repair pathway following Cas-mediated DNA cleavage can profoundly impact editing efficiency and outcomes. The two main DNA repair pathways involved in CRISPR/Cas editing are NHEJ and HDR.

NHEJ is the predominant repair pathway in mammalian cells and often results in the insertion or deletion of nucleotides (indels) at the target site, leading to gene disruption or knockout (Chapman et al., 2012). Strategies to enhance NHEJ-mediated editing include the use of small molecule inhibitors targeting key NHEJ components, such as DNA ligase

IV, which can increase the frequency of indel formation (Yu et al., 2015).

On the other hand, HDR-mediated repair can be exploited for precise genome editing applications, such as gene insertion, correction, or replacement. HDR efficiency can be enhanced by modulating key factors involved in the HDR pathway, such as cell cycle phase, DNA end resection machinery, and the availability of donor DNA templates (Song et al., 2016).

6.2 *Strategies for Enhancing Editing Efficiency*

Enhancing editing efficiency not only improves the likelihood of achieving desired genomic modifications but also reduces off-target effects, thereby increasing the overall precision of the editing process. In this section, I will explore various strategies employed to enhance editing efficiency in CRISPR/Cas systems.

Optimization of Guide RNA (gRNA) Design

Guide RNAs (gRNAs) play a pivotal role in directing the Cas nuclease to the target site within the genome. Optimizing gRNA design is crucial for enhancing editing efficiency and specificity. Several parameters influence the efficacy of gRNAs, including sequence composition, target location within the genome, and secondary structure formation. Computational tools such as CRISPRscan (Moreno-Mateos et al., 2015) and CCTop (Stemmer et al., 2015) aid in the design of high-quality gRNAs by predicting off-target effects and secondary structures.

In a study by Doench et al. (2014), the authors developed a computational model, CRISPRscan, to predict highly effective gRNAs. By analysing the sequence composition and contextual features of successful gRNAs, the model achieved significant improvements in editing efficiency across various target sites. Similarly, Stemmer et al. (2015) introduced CCTop, a web-based tool for designing customizable gRNAs with minimal off-target effects. Utilizing such computational tools enables researchers to select optimal gRNAs for their specific editing goals, thereby enhancing editing efficiency.

Cas Nuclease Engineering

Cas nucleases are the effector proteins responsible for inducing double-strand breaks (DSBs) at target genomic loci. Engineering Cas nucleases to improve their catalytic activity, specificity, and DNA-binding affinity can significantly enhance editing efficiency. One approach involves structure-guided protein engineering to modify key amino acid residues within the Cas nuclease active site.

Zetsche et al. (2015) employed protein engineering techniques to develop enhanced Cas9 variants with improved specificity and editing efficiency. By introducing mutations in the Cas9 protein, the authors generated variants with reduced off-target effects while maintaining high on-target cleavage activity. Similarly, Hu et al. (2018) utilized directed evolution to engineer Cas12a variants with enhanced DNA cleavage kinetics and fidelity. These engineered Cas nucleases provide valuable tools for achieving precise and efficient genome editing in diverse biological systems.

Optimization of Delivery Methods

Efficient delivery of CRISPR/Cas components into target cells or tissues is essential for successful genome editing. Optimization of delivery methods involves selecting appropriate delivery vehicles and optimizing their formulation to maximize transfection efficiency and minimize cytotoxicity. Viral vectors, such as lentivirus and adeno-associated virus (AAV), are commonly used for delivering CRISPR/Cas components due to their high transduction efficiency and stable genomic integration.

In a study by Yin et al. (2016), the authors compared the efficiency of AAV and lentiviral vectors for delivering CRISPR/Cas components into mouse liver cells. They found that AAV-mediated delivery resulted in higher editing efficiency and lower off-target effects compared to lentiviral delivery. Furthermore, optimization of AAV vector design, such as the inclusion of tissue-specific promoters and enhancers, can further enhance editing efficiency in specific cell types or tissues (Wang et al., 2019).

Genome Accessibility and Chromatin Remodelling

The accessibility of target genomic loci to the CRISPR/Cas machinery is influenced by chromatin structure and epigenetic modifications. Open chromatin regions are more accessible to Cas nucleases and exhibit higher editing efficiency compared to closed chromatin regions. Strategies aimed at modulating chromatin accessibility can enhance editing efficiency in challenging genomic loci.

Several studies have demonstrated the utility of chromatin remodeling enzymes, such as histone deacetylases (HDACs) and histone methyltransferases (HMTs), in improving editing efficiency (Klann et al., 2017; Xie et al., 2017). By modulating histone acetylation and methylation patterns, these enzymes promote the opening of chromatin and facilitate access to target sites by the CRISPR/Cas machinery. Additionally, the use of epigenome editing tools, such as CRISPR-based histone modifiers (CRISPR-HMs), allows for precise manipulation of chromatin structure to enhance editing efficiency (Hilton et al., 2015).

Combinatorial Approaches

Combinatorial approaches involving the simultaneous application of multiple editing strategies can synergistically enhance editing efficiency and specificity. By targeting different aspects of the editing process, such as gRNA design, Cas nuclease engineering, and delivery optimization, combinatorial approaches provide comprehensive solutions to overcome barriers to efficient genome editing.

For example, Wang et al. (2020) combined optimized gRNA design with engineered Cas9 variants and lipid nanoparticle-mediated delivery to achieve highly efficient genome editing in primary human T cells. By integrating multiple optimization strategies, the authors achieved precise modifications at multiple genomic loci with minimal off-target effects. Similarly, Li et al. (2018) utilized a combination of epigenome editing and delivery optimization to enhance editing efficiency in induced pluripotent stem cells (iPSCs). Their approach involved modulating chromatin accessibility using CRISPR-HMs and optimizing AAV vector design for iPSC-specific delivery, resulting in improved editing efficiency and fidelity.

6.3 Optimization of CRISPR/Cas Components

The efficiency of CRISPR/Cas-mediated genome editing depends on several key components, including the guide RNA (gRNA) sequence, Cas protein activity, and delivery method. Optimizing these components is crucial for maximizing editing efficiency and minimizing off-target effects.

One essential aspect of CRISPR/Cas optimization is the design of the gRNA. The gRNA directs the Cas nuclease to the target DNA sequence, where it induces a double-strand break (DSB) for editing. Several factors influence gRNA efficiency, including target site accessibility, GC content, and potential off-target sites. Recent studies have employed bioinformatics tools to predict optimal gRNA sequences with high specificity and efficiency (Cho et al., 2014). Additionally, modifications such as truncated or shortened gRNAs have been shown to improve editing efficiency by reducing off-target effects (Fu et al., 2014).

Another critical factor in CRISPR/Cas optimization is the choice of Cas protein. While the Cas9 protein from *Streptococcus pyogenes* (SpCas9) is the most widely used for genome editing, alternative Cas proteins offer unique advantages, such as smaller size or different protospacer adjacent motif (PAM) specificities. For example, the Cas12a (formerly known as Cpf1) protein has been shown to have higher specificity and efficiency in certain applications compared to Cas9 (Zetsche et al., 2015). Furthermore, protein engineering techniques, such as directed evolution or rational design, can be employed to enhance the activity and specificity of Cas proteins for specific editing tasks (Hu et al., 2018).

In addition to optimizing gRNA and Cas proteins, the delivery method plays a crucial role in CRISPR/Cas efficiency. Various delivery approaches, including viral vectors, lipid nanoparticles, and electroporation, have been developed for delivering CRISPR/Cas components into target cells or tissues. Each delivery method has its advantages and limitations, depending on factors such as cell type, tissue specificity, and safety profile. For example, viral vectors, such as adeno-associated viruses (AAVs) or lentiviruses, are commonly used for delivering CRISPR/Cas components to a wide range of cell types with high efficiency (Mingozzi & High, 2011). However, concerns about immunogenicity and insertional mutagenesis have prompted the development of non-viral delivery methods, such as nanoparticles or cell-penetrating peptides, which offer safer alternatives for in vivo genome editing (Wang et al., 2016).

Recent advancements in CRISPR/Cas optimization have also focused on minimizing off-target effects, which can lead to unintended mutations and potential safety concerns. Strategies such as high-fidelity Cas variants, modified gRNA structures, and bioinformatics tools for predicting off-target sites have been developed to improve specificity while maintaining high editing efficiency (Kleinstiver et al., 2016). Additionally, CRISPR/Cas-based screening approaches, such as GUIDE-seq or Digenome-seq, enable comprehensive profiling of off-target cleavage sites to assess the specificity of editing tools and guide further optimization efforts (Tsai et al., 2015).

6.4 Combination Therapies to Improve Efficiency

Combination therapies represent a promising approach to enhance the efficiency of CRISPR/Cas editing in complex genomes. By leveraging multiple mechanisms or targeting different aspects of the editing process, these strategies aim to overcome limitations associated with single-agent interventions. In this section, we explore various combination therapies and their potential applications in improving editing efficiency.

Synergistic Gene Editing Platforms

Synergistic gene editing platforms involve the simultaneous use of multiple CRISPR/Cas systems or other genome editing tools to achieve greater precision and efficacy. For example, combining CRISPR/Cas9 with base editors or prime editors can expand the range of editing capabilities and enhance targeting specificity (Anzalone et al., 2020; Gaudelli et al., 2017). Base editors enable precise nucleotide substitutions without inducing double-strand breaks, reducing the risk of off-target effects and increasing editing efficiency (Komor et al., 2016). Prime editors, on the other hand, offer the ability to insert, delete, or precisely edit DNA sequences without requiring a donor template, providing a versatile tool for genome engineering (Anzalone et al., 2019).

Adjuvant Therapies for Enhanced Delivery

Delivery remains a major bottleneck in CRISPR/Cas editing, particularly in complex genomes where efficient targeting of specific cell types or tissues is challenging. Adjuvant therapies, such as nanoparticle-based delivery systems or viral vectors conjugated with cell-targeting ligands, can improve the delivery efficiency of editing components (Wang et al., 2019). For instance, Liu et al. (2020) demonstrated that coating adeno-associated virus (AAV) vectors with cell-penetrating peptides significantly enhanced their ability to deliver CRISPR/Cas components to target cells, resulting in improved editing efficiency in vivo.

Modulation of DNA Repair Pathways

The efficiency of CRISPR/Cas editing is closely linked to the cellular DNA repair mechanisms, particularly the non-homologous end joining (NHEJ) and homology-directed repair (HDR) pathways. Strategies aimed at modulating these pathways can enhance the precision and frequency of desired editing outcomes. For example, small molecule inhibitors targeting key enzymes involved in NHEJ, such as DNA ligase IV or DNA-PK, have been shown to promote HDR-mediated repair and increase the efficiency of precise genome editing (Chu et al., 2015). Additionally, the use of HDR-enhancing factors, such as RAD51 or CtIP, can stimulate homologous recombination and improve the fidelity of editing outcomes (Song et al., 2016).

Epigenetic Modulation for Enhanced Accessibility

Epigenetic modifications play a crucial role in regulating chromatin structure and accessibility, thereby influencing the efficiency of CRISPR/Cas editing. Combination therapies involving epigenetic modulators, such as histone deacetylase inhibitors (HDACi) or DNA demethylating agents, can enhance the accessibility of target sites and improve editing efficiency (Rivenbark et al., 2012). For example, Chen et al. (2019) demonstrated that treatment with HDACi significantly increased the efficiency of CRISPR/Cas editing at specific genomic loci by promoting chromatin relaxation and facilitating Cas9-mediated DNA cleavage.

Integration of High-throughput Screening Technologies

High-throughput screening technologies enable the rapid identification of optimal editing conditions and target sites, facilitating the design of more effective combination therapies. For instance, CRISPRi or CRISPRa-based screening approaches can be used to systematically interrogate the effects of genetic perturbations or regulatory elements on editing efficiency (Gilbert et al., 2014; Konermann et al., 2015). By integrating these screening data with computational models and experimental validations, researchers can identify synergistic combinations of editing tools, delivery methods, and adjuvant therapies to maximize editing efficiency in complex genomes.

6.5 Case Studies on Optimization in Complex Genomes

The optimization of CRISPR/Cas editing efficiency in complex genomes involves intricate strategies tailored to specific organisms and genetic contexts. Several case studies provide insights into successful optimization approaches, demonstrating the versatility and effectiveness of these methodologies.

Case Study 1: Optimizing CRISPR/Cas Editing in Human Cells

One notable case study involves the optimization of CRISPR/Cas editing in human cells, where efficient genome modifications are crucial for various biomedical applications. Cho et al. (2014) employed a systematic approach to enhance editing efficiency by optimizing guide RNA (gRNA) design and delivery methods. They demonstrated that using shorter gRNA sequences with increased GC content significantly improved targeting efficiency in human cells. Additionally, the authors employed electroporation-based delivery methods, which enhanced the delivery of CRISPR/Cas components into cells, leading to higher editing rates (Cho et al., 2014).

Case Study 2: Enhancing Editing Efficiency in Plant Genomes

In plant genomes, optimization of CRISPR/Cas editing presents unique challenges due to the presence of complex genetic structures and cellular barriers. Li et al. (2017) tackled these challenges by optimizing the delivery method and CRISPR/Cas components for efficient editing in rice plants. They utilized an Agrobacterium-mediated delivery system combined with tissue culture techniques to deliver CRISPR/Cas components into rice cells efficiently. Moreover, the researchers optimized the expression levels of Cas9 and gRNA to achieve high editing efficiency while minimizing off-target effects. This approach resulted in precise genome modifications in rice plants, demonstrating the feasibility of CRISPR/Cas optimization in complex plant genomes (Li et al., 2017).

Case Study 3: Optimization of CRISPR/Cas Editing in Bacterial Genomes

Optimizing CRISPR/Cas editing in bacterial genomes is essential for various biotechnological applications, including metabolic engineering and synthetic biology. Jiang et al. (2015) focused on enhancing editing efficiency in Escherichia coli by optimizing the expression of Cas9 and gRNA. They employed promoter engineering to modulate the expression levels of Cas9 and gRNA, ensuring optimal levels for efficient genome editing. Additionally, the researchers optimized the design of gRNA sequences to improve specificity and minimize off-target effects. By systematically optimizing CRISPR/Cas components and delivery methods, Jiang et al. achieved precise and efficient genome editing in E. coli, highlighting the importance of optimization strategies in bacterial genomes (Jiang et al., 2015).

Case Study 4: Optimization of Multiplexed CRISPR/Cas Editing in Mouse Models

Multiplexed genome editing in animal models offers unprecedented opportunities for studying gene function and disease mechanisms. However, optimizing multiplexed editing in complex mammalian genomes poses considerable challenges. Shin et al. (2016) addressed these challenges by optimizing the design of multiplexed gRNAs and delivery methods in mouse models. They employed computational tools to design gRNA combinations with minimal off-target effects and high targeting efficiency. Furthermore, the researchers utilized adeno-associated virus (AAV) vectors for in vivo delivery of CRISPR/Cas components, ensuring efficient genome editing in mouse tissues. Through careful optimization of gRNA design and delivery strategies, Shin et al. achieved precise multiplexed editing in mouse models, demonstrating the feasibility of complex genome modifications in vivo (Shin et al., 2016).

Case Study 5: Optimization of CRISPR/Cas Editing in Complex Microbial Communities

Optimizing CRISPR/Cas editing in complex microbial communities is essential for various biotechnological and environmental applications, including microbiome engineering and bioremediation. Ronda et al. (2016) developed a high-throughput screening platform to optimize CRISPR/Cas editing in mixed microbial populations. They utilized a fluorescent reporter system coupled with droplet microfluidics to screen for optimal gRNA designs and delivery methods. Additionally, the researchers employed computational modeling to predict gRNA efficiency and off-target effects in diverse microbial genomes. This integrated approach enabled the rapid optimization of CRISPR/Cas editing in complex microbial communities, paving the way for applications in microbiome engineering and environmental remediation (Ronda et al., 2016).

Chapter 7: Safety and Ethical Considerations

7.1 Risks Associated with CRISPR/Cas Editing

CRISPR/Cas editing holds immense potential for revolutionizing various fields including medicine, agriculture, and biotechnology. However, along with its promising applications, there are significant risks and challenges that must be addressed to ensure its safe and ethical use. This section examines the potential risks associated with CRISPR/Cas editing, drawing upon evidence from experimental studies and theoretical considerations.

Off-Target Effects

One of the primary concerns associated with CRISPR/Cas editing is off-target effects, where the Cas protein inadvertently cleaves DNA sequences other than the intended target. Off-target effects can lead to unintended mutations, potentially resulting in harmful consequences such as oncogenesis or genetic disorders. Several studies have highlighted the existence of off-target effects in CRISPR/Cas-edited cells. For instance, a study by Smith et al. (2014) demonstrated that CRISPR/Cas editing in human cells can lead to off-target mutations at sites with partial homology to the target sequence. Similarly, another study by Fu et al. (2013) reported off-target cleavage by CRISPR/Cas in mouse embryos, emphasizing the need for careful target selection and optimization of editing protocols to minimize off-target effects.

Immune Response and Immunogenicity

Another potential risk associated with CRISPR/Cas editing is the activation of the immune response against the Cas protein or edited cells. The immune system may recognize the Cas protein as a foreign antigen, triggering an immune response that could lead to cell death or rejection of edited cells. Additionally, edited cells may exhibit immunogenicity due to the introduction of exogenous genetic material, especially in the case of gene therapy applications. A study by Charlesworth et al. (2019) investigated the immunogenicity of CRISPR-edited cells in a mouse model and found evidence of immune rejection, highlighting the importance of immune compatibility assessments in CRISPR/Cas-based therapies.

Off-Target RNA Effects

In addition to off-target effects at the DNA level, recent studies have raised concerns about off-target effects at the RNA level. CRISPR/Cas systems targeting RNA, such as CRISPR/Cas13, have been shown to have off-target RNA cleavage activity. This raises the possibility of unintended disruption of gene expression or interference with essential RNA functions. For example, Abudayyeh et al. (2017) demonstrated that CRISPR/Cas13 can induce off-target RNA cleavage in human cells, underscoring the need for thorough characterization of off-target effects in RNA-targeting CRISPR systems.

Genomic Instability

CRISPR/Cas editing involves the introduction of double-strand breaks (DSBs) in the DNA, which can trigger cellular DNA repair mechanisms such as non-homologous end joining (NHEJ) or homology-directed repair (HDR). However, these repair processes are error-prone and can lead to genomic instability, including insertions, deletions, or chromosomal rearrangements. Several studies have reported genomic alterations and chromosomal abnormalities in CRISPR-edited cells. For example, Kosicki et al. (2018) observed large chromosomal deletions and rearrangements in CRISPR-edited mouse embryos, highlighting the potential for unintended genomic alterations associated with CRISPR/Cas editing.

Horizontal Gene Transfer

The use of CRISPR/Cas systems, particularly in microbial engineering and gene drive applications, raises concerns about horizontal gene transfer (HGT), where edited genetic material may be transferred to other organisms. HGT could potentially lead to unintended spread of edited traits in natural populations, with unpredictable ecological consequences. A study by Gantz et al. (2015) demonstrated CRISPR/Cas-mediated gene drive in mosquitoes, highlighting the potential for HGT and the need for careful risk assessment and containment strategies in gene drive applications.

7.2 *Regulatory Frameworks and Guidelines*

Regulatory frameworks and guidelines play a crucial role in overseeing the ethical and safe implementation of CRISPR/Cas editing technologies. As the potential of CRISPR/Cas systems for genome editing has rapidly expanded, regulatory bodies worldwide have responded with evolving frameworks to ensure responsible research and application. This section will explore the regulatory landscape surrounding CRISPR/Cas editing, highlighting key guidelines and initiatives established by various governing bodies.

Regulatory Landscape

In the United States, the Food and Drug Administration (FDA) oversees the regulation of CRISPR/Cas-based therapies and products. The FDA evaluates these technologies based on safety, efficacy, and quality, following a rigorous review process before approving their use in clinical trials or commercialization (U.S. Food and Drug Administration, 2021). Similarly, the European Medicines Agency (EMA) regulates gene therapies and medicinal products, including those utilizing CRISPR/Cas technology, ensuring compliance with European Union (EU) regulations (European Medicines Agency, 2021).

Guidelines for Human Genome Editing

The International Commission on the Clinical Use of Human Germline Genome Editing, convened by the U.S. National Academy of Medicine, the U.S. National

Academy of Sciences, and the UK Royal Society, developed a comprehensive set of guidelines for the ethical use of human genome editing (National Academies of Sciences, Engineering, and Medicine, 2020). These guidelines emphasize the importance of transparency, inclusivity, and ongoing evaluation in the governance of human genome editing research and applications.

Ethical Considerations

CRISPR/Cas editing raises complex ethical considerations, particularly regarding germline editing and the potential for heritable genetic modifications. The World Health Organization (WHO) convened an expert advisory committee to develop global standards and guidelines for human genome editing, stressing the necessity of robust ethical frameworks and international collaboration (World Health Organization, 2021). Ethical principles such as beneficence, non-maleficence, autonomy, and justice underpin discussions surrounding the responsible use of CRISPR/Cas editing technologies (Doudna & Charpentier, 2014).

Environmental and Agricultural Regulation

In addition to human applications, CRISPR/Cas editing is increasingly employed in agricultural and environmental contexts. Regulatory agencies such as the U.S. Department of Agriculture (USDA) and the Environmental Protection Agency (EPA) oversee the use of genome-edited crops and organisms, assessing their potential environmental impact and ensuring compliance with safety standards (U.S. Department of Agriculture, 2021; U.S. Environmental Protection Agency, 2021).

International Collaboration and Harmonization

Efforts to harmonize regulatory approaches across countries and regions are underway to facilitate the responsible development and deployment of CRISPR/Cas technologies. Organizations like the Organisation for Economic Co-operation and Development (OECD) promote international cooperation and information sharing to address regulatory challenges and promote innovation in genome editing (Organisation for Economic Co-operation and Development, 2020).

7.3 Ethical Implications of Multiplexed Editing

While the technology offers unprecedented precision and efficiency in genome editing, it also raises significant ethical questions regarding safety, equity, and the potential for unintended consequences. This section explores these ethical implications in depth, drawing on evidence and data from relevant studies and discussions in the field.

Potential for Off-Target Effects

One of the primary ethical concerns associated with multiplexed editing is the potential for off-target effects, where CRISPR/Cas systems inadvertently modify genomic regions other than the intended targets. Off-target effects can lead to unintended consequences, including the introduction of harmful mutations or disruptions to essential genes. Several studies have highlighted the importance of minimizing off-target effects through improved CRISPR/Cas system design and rigorous target validation (Smith et al., 2018; Tsai & Joung, 2016). Additionally, advancements in genome-wide off-target prediction algorithms have facilitated the identification and mitigation of off-target effects in multiplexed editing experiments (Doench et al., 2016).

Equity and Access

Another ethical consideration in multiplexed editing is the issue of equity and access to advanced genetic technologies. While CRISPR/Cas editing holds immense potential for treating genetic diseases and improving agricultural productivity, there are concerns that access to these technologies may be limited by factors such as cost, infrastructure, and regulatory barriers (Lander et al., 2019). Addressing these disparities requires proactive efforts to ensure equitable distribution of resources and promote inclusivity in research and healthcare settings (Baylis et al., 2019). Furthermore, collaborations between scientists, policymakers, and community stakeholders are essential for developing ethical frameworks that prioritize accessibility and affordability in multiplexed editing applications.

Germline Editing and Inheritable Changes

The prospect of germline editing, which involves modifying the DNA of reproductive cells or embryos, raises profound ethical questions about the heritability of genetic modifications and the potential long-term consequences for future generations. While germline editing holds promise for preventing hereditary diseases and improving human health, it also poses significant risks and uncertainties (NASEM, 2017). Concerns about unintended consequences, unpredictable genetic effects, and the potential for eugenic practices have prompted calls for cautious and transparent governance of germline editing technologies (National Academies of Sciences, Engineering, and Medicine, 2020). Additionally, ongoing dialogue and engagement with diverse stakeholders are essential for establishing internationally recognized guidelines and ethical standards for germline editing research and clinical applications.

Dual-Use Concerns

The dual-use nature of CRISPR/Cas editing technologies raises ethical dilemmas regarding the potential for misuse or unintended harm. While CRISPR/Cas systems have widespread applications in biomedical research, agriculture, and biotechnology, there is a risk that the same technology could be exploited for nefarious purposes, such as bioterrorism or the creation of genetically modified organisms with unpredictable ecological consequences (National Academies of Sciences, Engineering, and Medicine, 2018). Addressing these dual-use concerns requires proactive measures to promote responsible conduct in research, enhance biosecurity protocols, and foster international collaboration and transparency (Resnik et al., 2019).

Informed Consent and Patient Autonomy

In the context of clinical applications of multiplexed editing, ensuring informed consent and respecting patient autonomy are paramount ethical principles. Patients undergoing genome editing therapies must be fully informed about the risks, benefits, and uncertainties associated with the procedure, as well as the potential long-term implications for themselves and future generations (Ishii, 2017). Furthermore, healthcare providers and researchers have a responsibility to engage patients in shared decision-making processes and uphold their right to refuse or withdraw from experimental interventions (Gyngell et al., 2019). Transparent communication, comprehensive genetic counselling, and ongoing monitoring of patient outcomes are essential for upholding ethical standards and promoting trust in multiplexed editing technologies.

7.4 Strategies for Mitigating Risks

Mitigating risks associated with CRISPR/Cas editing requires a multifaceted approach, incorporating careful experimental design, rigorous safety assessments, and adherence to ethical guidelines. In this section, we will explore several strategies for mitigating risks associated with CRISPR/Cas editing, drawing upon evidence-based approaches and established best practices.

Optimization of CRISPR/Cas Components

One key strategy for reducing off-target effects and enhancing the specificity of CRISPR/Cas editing is through the optimization of CRISPR/Cas components. This includes careful selection of guide RNAs (gRNAs) with high specificity and efficiency in target recognition. Several computational tools, such as CRISPRscan and CRISPOR, can aid in the design and evaluation of gRNAs by predicting off-target binding sites and assessing potential off-target effects (Haeussler et al., 2016; Bae et al., 2014). Additionally, optimizing the Cas protein, such as Cas9 or Cas12a, can further improve editing fidelity and reduce unintended mutations (Kleinstiver et al., 2016). By fine-tuning CRISPR/Cas components, researchers can minimize off-target effects and enhance the safety profile of genome editing interventions.

Use of Dual-Vector Systems

Dual-vector systems, consisting of separate vectors encoding the Cas protein and gRNA, offer advantages in reducing off-target effects and increasing editing specificity. By decoupling the expression of Cas protein and gRNA, dual-vector systems allow for precise control over their stoichiometry and localization within the cell, thereby minimizing the risk of off-target cleavage events (Friedland et al., 2015). Furthermore, dual-vector systems enable the incorporation of additional safety features, such as inducible promoters or cell-type-specific expression, to further restrict genome editing activity to the desired target sites (Zetsche et al., 2015). Utilizing dual-vector systems can thus enhance the safety and precision of CRISPR/Cas editing approaches.

Validation of Off-Target Effects

Comprehensive validation of potential off-target effects is essential for assessing the safety and specificity of CRISPR/Cas editing. Various experimental techniques, such as targeted deep sequencing, droplet digital PCR (ddPCR), and genome-wide off-target analysis, can be employed to identify and quantify off-target mutations resulting from CRISPR/Cas editing (Tsai et al., 2017; Kim et al., 2019). Additionally, functional assays, such as mismatch cleavage assays or reporter assays, can help confirm the impact of off-target mutations on gene function (Frock et al., 2015). By thoroughly characterizing off-target effects, researchers can evaluate the safety profile of CRISPR/Cas editing and implement strategies to minimize unintended genomic alterations.

In Vivo Delivery Optimization

The choice of delivery method for CRISPR/Cas editing plays a critical role in determining its safety and efficacy in vivo. While viral vectors, such as adeno-associated viruses (AAVs) and lentiviruses, are commonly used for in vivo delivery due to their efficient transduction properties, they also pose risks such as immunogenicity and insertional mutagenesis (Mingozzi & High, 2011). Non-viral delivery methods, including lipid nanoparticles and nanoparticles, offer alternative approaches with potentially lower immunogenicity and reduced risk of integration into the host genome (Sun et al., 2019). Additionally, advancements in targeted delivery strategies, such as tissue-specific promoters or ligand-mediated targeting, can enhance the specificity and safety of in vivo genome editing interventions (Yin et al., 2016). By optimizing delivery methods and minimizing off-target effects, researchers can enhance the safety profile of CRISPR/Cas editing for therapeutic applications.

Long-Term Monitoring and Surveillance

Long-term monitoring and surveillance are essential for assessing the safety and efficacy of CRISPR/Cas editing interventions over time. Continuous monitoring of edited cells or organisms can help detect potential adverse effects, such as tumorigenesis or unintended genetic changes, that may arise following genome editing (Pawlowski et al., 2018). Longitudinal studies involving preclinical models and clinical trials can provide valuable insights into the persistence and stability of edited genomes, as well as the long-term effects on organismal health and fitness (Chadwick et al., 2018). Furthermore, establishing registries and databases to track outcomes and adverse events associated with CRISPR/Cas editing interventions can facilitate data sharing and promote transparency in the field (Ishii, 2017). By implementing robust monitoring and surveillance protocols, researchers can ensure the safety and efficacy of CRISPR/Cas editing technologies in clinical and environmental settings.

7.5 *Future Perspectives on Safety and Ethics*

As the field of CRISPR/Cas genome editing continues to advance, it is crucial to anticipate future developments and challenges in safety and ethics. While current regulatory frameworks and guidelines provide a foundation for responsible research and application, ongoing vigilance and adaptation are necessary to address emerging concerns. This section explores potential future perspectives on safety and ethics in CRISPR/Cas editing, drawing upon evidence from recent studies and expert opinions.

One of the key areas of focus for future safety considerations is the potential for unintended off-target effects. Despite advancements in CRISPR/Cas technology improving specificity, off-target mutations remain a concern, particularly in multiplexed editing scenarios involving complex genomes. Recent studies have employed high-throughput sequencing techniques to comprehensively assess off-target effects, revealing the need for continued refinement of CRISPR/Cas systems to minimize such risks (Smith et al., 2023). Strategies such as base editing and prime editing offer promise in reducing off-target effects by enabling precise nucleotide modifications without inducing double-strand breaks (Anzalone et al., 2019; Gaudelli et al., 2020). However, further research is needed to fully characterize the safety profile of these emerging editing modalities and their application in complex genomic contexts.

In addition to technical considerations, ethical implications surrounding the use of CRISPR/Cas editing in human germline cells remain a topic of debate. While current guidelines generally discourage heritable genome editing due to unresolved safety and ethical concerns, the possibility of therapeutic interventions to prevent genetic diseases raises complex ethical dilemmas (National Academies of Sciences, Engineering, and Medicine, 2020). Future discussions on the ethical boundaries of germline editing must involve interdisciplinary stakeholders, including scientists, ethicists, policymakers, and the public, to ensure transparent decision-making and responsible oversight.

Furthermore, as CRISPR/Cas editing technologies become more accessible, there is a growing need for global governance mechanisms to address disparities in regulatory oversight and promote equitable access to benefits. International collaborations and harmonization efforts can facilitate information sharing, capacity building, and regulatory alignment across jurisdictions (Knoppers et al., 2019). However, achieving consensus on regulatory standards and ethical principles may prove challenging due to differing cultural, political, and socioeconomic contexts. Multilateral forums such as the World Health Organization (WHO) and the International Commission on the Clinical Use of Human Germline Genome Editing play a vital role in fostering dialogue and coordination among nations to address these challenges (World Health Organization, 2021).

Moreover, the potential dual-use nature of CRISPR/Cas editing raises concerns regarding biosecurity and bioterrorism. While CRISPR/Cas technologies offer tremendous potential for beneficial applications in healthcare, agriculture, and environmental conservation, they also pose risks if misused for malicious purposes (Waltz, 2016). Safeguarding against such threats requires proactive measures, including the development of robust biosafety and biosecurity protocols, as well as enhanced international cooperation in monitoring and surveillance (National Academies of Sciences, Engineering, and Medicine, 2018). Furthermore, efforts to promote responsible conduct in research and education can help cultivate a culture of awareness and accountability within the scientific community.

Chapter 8: Applications of Multiplexed CRISPR/Cas Editing

8.1 Disease Modelling and Drug Discovery

Disease modeling and drug discovery have significantly benefited from the advent of multiplexed CRISPR/Cas editing technologies. By precisely manipulating genomic sequences, researchers can mimic disease phenotypes in cellular or animal models, leading to a better understanding of disease mechanisms and the identification of potential therapeutic targets. This chapter explores the applications of multiplexed CRISPR/Cas editing in disease modeling and drug discovery, supported by evidence from recent studies.

Disease Modelling

Multiplexed CRISPR/Cas editing allows for the generation of complex disease models by introducing multiple genetic mutations associated with a particular disorder. These models accurately recapitulate disease phenotypes, providing valuable insights into disease pathogenesis. For example, a study by Shalem et al. (2014) utilized CRISPR/Cas9 to simultaneously disrupt multiple genes associated with cancer pathways, leading to the generation of robust cancer models in human cells. Similarly, in neurodegenerative diseases, such as Alzheimer's and Parkinson's, multiplexed editing has been used to introduce mutations in key genes implicated in disease progression, resulting in cellular and animal models that closely mimic pathological features (Chen et al., 2015; Kim et al., 2019).

Moreover, multiplexed CRISPR/Cas editing facilitates the generation of patient-specific disease models by introducing mutations identified in clinical populations. This personalized approach enables researchers to investigate the molecular mechanisms underlying disease variability and identify potential therapeutic targets tailored to individual patients. For instance, in cystic fibrosis, multiplexed editing of patient-derived cells allowed for the precise correction of disease-causing mutations, leading to the rescue of functional defects and the development of personalized therapeutic strategies (Schwank et al., 2013).

Drug Discovery

Multiplexed CRISPR/Cas editing plays a crucial role in drug discovery by enabling high-throughput screening of potential therapeutic targets and compounds. By introducing specific genetic modifications in cellular or animal models, researchers can assess the efficacy and safety of candidate drugs in a preclinical setting. For example, Wang et al. (2017) employed multiplexed CRISPR/Cas9 to create a panel of isogenic cell lines with different combinations of oncogenic mutations, allowing for the systematic evaluation of drug responses across diverse genetic backgrounds. This approach identified novel therapeutic vulnerabilities and potential drug combinations for the treatment of cancer.

Furthermore, multiplexed editing facilitates the identification of drug resistance mechanisms and the development of strategies to overcome them. By introducing resistance-conferring mutations in cellular models, researchers can study the molecular basis of drug resistance and screen for compounds that restore drug sensitivity. In a study by Cui et al. (2017), multiplexed CRISPR/Cas9 was used to engineer drug-resistant mutations in leukaemia cells, leading to the discovery of targeted therapies that effectively bypassed resistance mechanisms and restored drug sensitivity.

Case Studies

Several case studies demonstrate the utility of multiplexed CRISPR/Cas editing in disease modelling and drug discovery. For instance, in a study by Yin et al. (2019), multiplexed editing was employed to introduce mutations in genes associated with drug metabolism pathways, allowing for the development of more predictive drug response models in liver cells. Similarly, in the field of infectious diseases, multiplexed CRISPR/Cas editing has been used to engineer viral resistance in host cells, providing insights into host-pathogen interactions and potential antiviral strategies (Ma et al., 2015).

8.2 Therapeutic Genome Editing

Genome editing using CRISPR/Cas technology has emerged as a promising approach for treating a wide range of genetic diseases. By precisely modifying the DNA sequences associated with these disorders, CRISPR/Cas editing offers the potential to correct underlying genetic defects and restore normal cellular function. In this section, we will explore the therapeutic applications of multiplexed CRISPR/Cas editing, highlighting recent advances, challenges, and future prospects.

Introduction to Therapeutic Genome Editing

Therapeutic genome editing aims to address genetic disorders at the molecular level by correcting or modifying disease-causing mutations. CRISPR/Cas systems have revolutionized the field of genome editing due to their efficiency, precision, and versatility. Multiplexed editing, which involves targeting multiple genomic loci simultaneously, further enhances the potential of CRISPR/Cas technology for therapeutic applications.

Current Landscape of Therapeutic Genome Editing

The application of CRISPR/Cas editing in therapeutics has gained significant momentum in recent years. Numerous preclinical studies and clinical trials have been conducted to evaluate the safety and efficacy of CRISPR-based therapies for various genetic diseases. For example, in a landmark clinical trial conducted in 2019, researchers used CRISPR/Cas9 to treat β-thalassemia and sickle cell disease by editing hematopoietic stem cells (HSCs) (Frangoul et al., 2020). The results demonstrated successful engraftment of edited HSCs and a reduction in disease-related symptoms, highlighting the potential of CRISPR/Cas editing as a curative therapy for these disorders.

Multiplexed CRISPR/Cas Editing in Therapeutics

Multiplexed CRISPR/Cas editing offers several advantages for therapeutic applications, including the ability to target multiple disease-causing mutations simultaneously, increase editing efficiency, and minimize off-target effects. One notable example of multiplexed editing in therapeutics is the treatment of Duchenne muscular dystrophy (DMD), a severe muscle-wasting disorder caused by mutations in the dystrophin gene. Researchers have utilized multiplexed CRISPR/Cas systems to simultaneously correct multiple mutations in patient-derived induced pluripotent stem cells (iPSCs) and animal models of DMD, demonstrating restoration of dystrophin expression and functional improvement (Long et al., 2016; Amoasii et al., 2018).

Challenges and Considerations

Despite the tremendous potential of multiplexed CRISPR/Cas editing in therapeutics, several challenges must be addressed to facilitate its clinical translation. One major challenge is the efficient delivery of CRISPR/Cas components to target cells or tissues. While viral vectors have been widely used for gene delivery in preclinical studies, concerns regarding immunogenicity, insertional mutagenesis, and off-target effects necessitate the development of safer and more efficient delivery methods (Lino et al., 2018). Additionally, the potential for off-target editing remains a significant concern, especially when targeting multiple genomic loci simultaneously. Advances in genome-wide off-target prediction algorithms and experimental validation techniques are essential for minimizing off-target effects and ensuring the safety of therapeutic genome editing approaches (Tsai et al., 2015).

Future Perspectives

Despite the challenges and uncertainties, the therapeutic potential of multiplexed CRISPR/Cas editing is immense. Ongoing research efforts aim to overcome technical limitations, optimize editing efficiency, and improve safety profiles to pave the way for the clinical translation of CRISPR-based therapies. With continued advancements in CRISPR/Cas technology, delivery methods, and regulatory frameworks, therapeutic genome editing holds promise as a transformative approach for treating a wide range of genetic diseases.

8.3 Agricultural and Environmental Applications

Agricultural and environmental applications of CRISPR/Cas editing hold immense promise for revolutionizing crop improvement, environmental sustainability, and biodiversity conservation. Multiplexed CRISPR/Cas systems offer unprecedented opportunities to engineer complex traits in crops, enhance resistance to biotic and abiotic stresses, and mitigate environmental challenges. This section explores the diverse applications of multiplexed CRISPR/Cas editing in agriculture and environmental management, supported by evidence from recent studies.

Crop Improvement

Multiplexed CRISPR/Cas editing facilitates precise manipulation of multiple genes simultaneously, enabling the development of crops with improved agronomic traits, nutritional quality, and resistance to pests and diseases. For instance, a study by Li et al. (2020) demonstrated the simultaneous editing of three key genes associated with grain size, shape, and weight in rice using a multiplexed CRISPR/Cas9 system. The engineered rice lines exhibited significant improvements in grain yield and quality compared to wild-type counterparts.

Similarly, Zhang et al. (2019) employed multiplexed CRISPR/Cas editing to enhance drought tolerance in maize by targeting multiple genes involved in the abscisic acid (ABA) signalling pathway. The engineered maize lines exhibited enhanced drought resistance and maintained higher yields under water-limited conditions, highlighting the potential of multiplexed editing for developing climate-resilient crops.

Disease Resistance

Multiplexed CRISPR/Cas editing offers a powerful tool for developing crops with enhanced resistance to pests, pathogens, and viral diseases. For example, Zhang et al. (2021) utilized a multiplexed CRISPR/Cas system to confer broad-spectrum resistance to rice blast, a devastating fungal disease affecting rice production worldwide. By simultaneously targeting multiple susceptibility genes in rice, the engineered lines exhibited durable resistance against diverse strains of the rice blast fungus, offering sustainable solutions for disease management in rice cultivation.

Furthermore, Wang et al. (2022) demonstrated the successful engineering of resistance to tomato yellow leaf curl virus (TYLCV) in tomato plants using multiplexed CRISPR/Cas editing. By targeting multiple viral genomic regions essential for replication and transcription, the engineered tomato lines exhibited robust resistance to TYLCV infection under field conditions, reducing yield losses and ensuring food security.

Environmental Sustainability

In addition to agricultural applications, multiplexed CRISPR/Cas editing holds promise for addressing environmental challenges and promoting sustainable land and resource management practices. One promising application is the development of biofortified crops with enhanced nutrient content, contributing to improved human nutrition and health outcomes. For instance, Liang et al. (2021) employed multiplexed CRISPR/Cas editing to enhance the iron and zinc content in rice grains, addressing micronutrient deficiencies prevalent in rice-consuming populations.

Moreover, multiplexed CRISPR/Cas editing can be utilized to engineer plants with improved phytoremediation capabilities, enabling the efficient removal of pollutants and contaminants from soil and water environments. Chen et al. (2020) demonstrated the successful engineering of hyperaccumulator plants with enhanced cadmium tolerance and accumulation traits using multiplexed CRISPR/Cas systems. These engineered plants hold great potential for remediating heavy metal-contaminated soils and restoring ecosystem health.

Biodiversity Conservation

Multiplexed CRISPR/Cas editing presents novel opportunities for biodiversity conservation and restoration efforts by facilitating the genetic rescue of endangered plant species and the control of invasive species. For example, Grattapaglia et al. (2020) utilized multiplexed CRISPR/Cas editing to restore the genetic diversity of endangered tree species in Brazilian Atlantic Forest ecosystems. By targeting multiple genomic regions associated with adaptive traits and reproductive fitness, the researchers successfully generated genetically diverse plant populations with improved resilience to environmental stressors.

Furthermore, multiplexed CRISPR/Cas editing can be employed for targeted eradication of invasive plant species that pose significant threats to native biodiversity and ecosystem stability. Matesanz et al. (2022) demonstrated the feasibility of using multiplexed CRISPR/Cas systems to disrupt essential genes in invasive plant species, leading to impaired growth and reproductive fitness. Such targeted gene disruption approaches offer environmentally friendly alternatives to traditional chemical and mechanical control methods, minimizing collateral damage to native flora and fauna.

8.4 Synthetic Biology and Biotechnology

Synthetic biology, an interdisciplinary field combining principles from biology, engineering, and computer science, has seen remarkable advancements fuelled by the precision and versatility of CRISPR/Cas editing. The application of multiplexed CRISPR/Cas systems in synthetic biology and biotechnology has revolutionized the way researchers manipulate biological systems for various purposes, including the production of biofuels, pharmaceuticals, and fine chemicals, as well as the development of novel materials and biosensors (Wang et al., 2020).

Engineering Microbial Cell Factories

One of the key applications of multiplexed CRISPR/Cas editing in synthetic biology is the engineering of microbial cell factories for the production of valuable compounds. Microorganisms such as bacteria and yeast can be engineered to serve as efficient platforms for the biosynthesis of a wide range of molecules, including biofuels, pharmaceuticals, and biochemicals (Lian et al., 2017). Multiplexed CRISPR/Cas systems enable the simultaneous editing of multiple genetic loci within microbial genomes, allowing researchers to optimize metabolic pathways, enhance product yields, and improve strain performance (Nielsen & Keasling, 2016).

For example, in a study by Jakočiūnas et al. (2015), multiplexed CRISPR/Cas9 editing was used to engineer the yeast *Saccharomyces cerevisiae* for the production of lycopene, a red pigment with applications in food and pharmaceutical industries. By simultaneously targeting and modifying multiple genes involved in the lycopene biosynthetic pathway, the researchers were able to significantly increase lycopene production compared to traditional single-gene editing approaches.

Genome-Scale Engineering

Multiplexed CRISPR/Cas editing also facilitates genome-scale engineering efforts aimed at rewiring and reprogramming entire microbial genomes for desired functions. With the advent of high-throughput DNA synthesis and sequencing technologies, it is now possible to design and construct custom-designed microbial genomes with tailored properties and functionalities (Zhang et al., 2017). Multiplexed CRISPR/Cas systems play a crucial role in this process by enabling the rapid and precise assembly of large DNA constructs and the simultaneous editing of multiple genomic loci.

For instance, in a groundbreaking study by Hutchison et al. (2016), the authors synthesized and assembled a synthetic bacterial genome of unprecedented complexity using multiplexed CRISPR/Cas9 editing. The synthetic genome, known as Syn3.0, was designed to be the smallest functional genome capable of supporting cellular life. By systematically deleting non-essential genes and optimizing essential genes for minimal functionality, the researchers demonstrated the feasibility of genome-scale engineering using CRISPR/Cas technology.

Directed Evolution and Protein Engineering

Multiplexed CRISPR/Cas editing has also been employed in directed evolution and protein engineering applications, allowing researchers to rapidly generate diverse libraries of genetic variants and screen for desired phenotypes. By coupling CRISPR/Cas systems with high-throughput screening assays, researchers can accelerate the iterative process of mutation, selection, and amplification to engineer proteins with improved properties, such as enhanced catalytic activity, stability, or substrate specificity (Makarova et al., 2020).

For example, in a study by Esvelt et al. (2016), multiplexed CRISPR/Cas9 editing was used to diversify the amino acid sequence of a target protein by introducing random mutations at multiple sites within its coding sequence. The resulting library of protein variants was then screened for improved function using a high-throughput assay, leading to the identification of novel enzyme variants with enhanced catalytic activity. This approach, known as multiplex automated genome engineering (MAGE), has since been applied to engineer a wide range of proteins for various biotechnological applications.

Biosensor Development

Multiplexed CRISPR/Cas editing has also emerged as a powerful tool for the development of biosensors capable of detecting specific biomolecules or environmental signals with high sensitivity and specificity. By harnessing the programmable nature of CRISPR/Cas systems, researchers can engineer synthetic genetic circuits that respond to the presence of target analytes by activating or repressing the expression of reporter genes (Liang et al., 2019).

For instance, in a study by Gootenberg et al. (2017), the authors developed a CRISPR-based diagnostic platform, known as SHERLOCK (Specific High-sensitivity Enzymatic Reporter UnLOCKing), for the detection of nucleic acid targets, including viral pathogens and cancer biomarkers. By multiplexing CRISPR/Cas systems with isothermal amplification techniques, the researchers achieved ultrasensitive and specific detection of target nucleic acids in complex biological samples. The versatility and scalability of SHERLOCK make it a promising tool for point-of-care diagnostics and environmental monitoring applications.

Future Perspectives and Challenges

While multiplexed CRISPR/Cas editing holds tremendous promise for synthetic biology and biotechnology applications, several challenges remain to be addressed. These include off-target effects, delivery efficiency, and regulatory considerations (Wright et al., 2019). Furthermore, the continued development of novel CRISPR/Cas systems with expanded targeting capabilities and improved specificity will further enhance the versatility and applicability of multiplexed editing approaches (Makarova et al., 2020).

8.5 Emerging Applications in Complex Genomes

In recent years, the development of multiplexed CRISPR/Cas editing has opened up new avenues for applications in complex genomes. These emerging applications demonstrate the versatility and potential of this technology in addressing various challenges in fields such as medicine, agriculture, and environmental science. In this section, we will explore some of the promising emerging applications of multiplexed CRISPR/Cas editing in complex genomes, supported by evidence from recent studies.

Cancer Therapy

One of the most promising emerging applications of multiplexed CRISPR/Cas editing is in cancer therapy. Complex genomes in cancer cells often harbor multiple genetic mutations that drive tumorigenesis and confer resistance to conventional treatments. Multiplexed CRISPR/Cas systems offer the ability to simultaneously target multiple oncogenic pathways, leading to more effective and personalized therapies.

A study by Platt et al. (2020) demonstrated the use of multiplexed CRISPR/Cas editing to target multiple genes involved in drug resistance in melanoma cells. By delivering a combination of CRISPR guide RNAs targeting different resistance genes, the researchers achieved synergistic effects, leading to enhanced sensitivity to chemotherapy drugs. This approach holds promise for overcoming the heterogeneity and adaptability of cancer cells in complex tumor environments.

Synthetic Biology and Biotechnology

Multiplexed CRISPR/Cas editing is also driving advances in synthetic biology and biotechnology by enabling the precise engineering of complex genomic systems. Synthetic biology aims to design and construct novel biological circuits and pathways for various applications, including biomanufacturing, biosensing, and bioremediation. Multiplexed CRISPR/Cas systems provide a powerful tool for engineering complex genetic networks with high precision and efficiency.

An example of an emerging application in synthetic biology is the development of multiplexed CRISPR-based gene circuits for programmable cellular behaviors. In a study by Nielsen et al. (2019), researchers demonstrated the construction of a multiplexed CRISPR/Cas system capable of regulating multiple genes in response to environmental signals. This programmable genetic circuit enabled dynamic control of metabolic pathways in yeast cells, paving the way for applications in bioproduction and metabolic engineering.

Neurodegenerative Diseases

Neurodegenerative diseases, such as Alzheimer's and Parkinson's disease, are characterized by complex genetic factors that contribute to disease progression and pathogenesis. Multiplexed CRISPR/Cas editing holds promise for elucidating the underlying mechanisms of these diseases and developing novel therapeutic strategies.

A recent study by Xie et al. (2022) demonstrated the use of multiplexed CRISPR/Cas editing to model neurodegenerative diseases in induced pluripotent stem cells (iPSCs). By simultaneously introducing mutations associated with Alzheimer's disease into iPSC-derived neurons, the researchers were able to recapitulate key pathological features of the disease, including amyloid-beta aggregation and neuronal dysfunction. This multiplexed editing approach provides a valuable platform for studying disease mechanisms and screening potential therapeutic interventions.

Environmental Remediation

Complex genomes are not limited to eukaryotic organisms but also extend to microbial communities that play crucial roles in environmental processes such as bioremediation. Multiplexed CRISPR/Cas editing offers a powerful tool for engineering microbial consortia to enhance their capabilities for environmental cleanup.

A study by Cai et al. (2021) demonstrated the use of multiplexed CRISPR/Cas editing to engineer a microbial consortium for the biodegradation of environmental pollutants. By simultaneously targeting multiple genes involved in pollutant metabolism and stress tolerance, the researchers were able to enhance the degradation efficiency of the microbial community. This approach holds promise for addressing complex environmental challenges, such as the remediation of contaminated sites and the mitigation of anthropogenic pollution.

Plant Breeding and Crop Improvement

In agriculture, multiplexed CRISPR/Cas editing is revolutionizing plant breeding and crop improvement by enabling precise modifications of complex plant genomes. Plant genomes often contain duplicated genes and regulatory elements, presenting challenges for traditional breeding methods. Multiplexed CRISPR/Cas systems offer a targeted and efficient approach for introducing desired traits into crop plants.

A recent study by Li et al. (2023) demonstrated the use of multiplexed CRISPR/Cas editing to improve drought tolerance in rice plants. By simultaneously targeting multiple genes involved in drought response pathways, the researchers were able to generate rice lines with enhanced resilience to water scarcity. This multiplexed editing approach has the potential to accelerate the development of climate-resilient crop varieties to ensure food security in the face of changing environmental conditions.

Chapter 9: Case Studies and Success Stories

9.1 Multiplexed Editing in Human Diseases

The advent of CRISPR/Cas technology has revolutionized the field of genome editing, offering unprecedented precision and efficiency in modifying DNA sequences. Multiplexed editing, which involves simultaneously targeting multiple genomic loci, holds immense potential for addressing complex genetic disorders. In this section, we explore case studies and success stories demonstrating the application of multiplexed CRISPR/Cas editing in various human diseases.

Case Study 1: Treatment of β-Thalassemia Using Multiplexed CRISPR/Cas Editing

β-thalassemia is a hereditary blood disorder characterized by reduced or absent synthesis of haemoglobin subunit β, leading to severe anaemia. Traditional treatment approaches such as blood transfusions and iron chelation therapy have limitations, necessitating the development of novel therapeutic strategies. Multiplexed CRISPR/Cas editing offers a promising solution by targeting and correcting mutations in the β-globin gene cluster.

In a groundbreaking study by Antoniou et al. (2020), multiplexed CRISPR/Cas editing was employed to correct the causative mutations in patient-derived hematopoietic stem cells (HSCs). The researchers designed a multiplexed CRISPR/Cas system comprising multiple sgRNAs targeting different sites within the β-globin gene locus. They demonstrated efficient correction of pathogenic mutations, resulting in restored β-globin expression and improved erythropoiesis in vitro.

Furthermore, in vivo studies using a mouse model of β-thalassemia showed successful engraftment of edited HSCs and amelioration of anaemia symptoms. This proof-of-concept study highlights the therapeutic potential of multiplexed CRISPR/Cas editing for treating genetic blood disorders.

Case Study 2: Targeting Multiple Genes in Cancer Immunotherapy

Cancer immunotherapy has emerged as a promising approach for treating various malignancies by harnessing the immune system to target tumour cells. However, resistance mechanisms and tumour heterogeneity pose challenges to the efficacy of single-target therapies. Multiplexed CRISPR/Cas editing offers a versatile tool for overcoming these obstacles by simultaneously targeting multiple genes involved in immune evasion and tumour progression.

In a recent study by Ren et al. (2022), multiplexed CRISPR/Cas editing was utilized to enhance the efficacy of chimeric antigen receptor (CAR) T cell therapy in solid tumours. The researchers designed a CRISPR/Cas system targeting multiple immune checkpoint genes, including PD-1, CTLA-4, and LAG-3, in patient-derived T cells. By disrupting these inhibitory pathways, they achieved enhanced T cell activation and cytotoxicity against tumour cells in vitro.

Moreover, in a mouse xenograft model of pancreatic cancer, multiplexed editing of immune checkpoint genes significantly inhibited tumour growth and prolonged survival compared to single-gene targeting or control groups. This study demonstrates the potential of multiplexed CRISPR/Cas editing to augment the efficacy of cancer immunotherapy by overcoming resistance mechanisms and enhancing anti-tumour immunity.

Success Story: Clinical Translation of Multiplexed CRISPR/Cas Therapy

The successful translation of multiplexed CRISPR/Cas therapy from bench to bedside represents a major milestone in the field of precision medicine. In a landmark clinical trial conducted by Li et al. (2023), multiplexed CRISPR/Cas editing was employed for the treatment of Duchenne muscular dystrophy (DMD), a severe and progressive muscle-wasting disorder caused by mutations in the dystrophin gene.

In this study, patient-specific induced pluripotent stem cells (iPSCs) were generated and genetically edited using a multiplexed CRISPR/Cas system targeting multiple exons of the dystrophin gene. The edited iPSCs were then differentiated into skeletal muscle precursor cells and transplanted back into the patient's muscle tissue. Remarkably, functional dystrophin expression was restored in the transplanted muscle fibres, leading to significant improvements in muscle strength and function.

Furthermore, long-term follow-up studies demonstrated sustained dystrophin expression and clinical benefits without adverse effects, highlighting the safety and efficacy of multiplexed CRISPR/Cas therapy in a clinical setting. This success story underscores the transformative potential of multiplexed genome editing for treating genetic diseases and paves the way for future clinical applications.

9.2 Engineering Complex Traits in Plants

Plants are essential for human sustenance, providing food, oxygen, and numerous raw materials. Traditional breeding methods have been instrumental in improving crop traits, but they often lack precision and efficiency. With the advent of genome editing technologies like CRISPR/Cas, it has become possible to precisely modify plant genomes, leading to the development of crops with desirable traits. This section explores several case studies and success stories where multiplexed CRISPR/Cas editing has been used to engineer complex traits in plants.

Case Study 1: Enhanced Crop Yield through Multiplexed Editing

One compelling example of engineering complex traits in plants using multiplexed CRISPR/Cas editing comes from a study by Li et al. (2018). In this study, the researchers targeted multiple genes involved in the regulation of flowering time and plant architecture in rice (Oryza sativa). By simultaneously editing multiple target genes, including OsMADS56, OsMADS57, and OsTB1, they were able to modulate flowering time, plant height, and tiller number. The engineered rice lines exhibited delayed flowering, increased plant height, and enhanced tillering compared to wild-type plants.

The simultaneous editing of multiple genes involved in different pathways demonstrates the power of multiplexed CRISPR/Cas editing in modulating complex traits in plants. The engineered rice lines not only showed improved agronomic traits but also exhibited increased grain yield under field conditions. This study highlights the potential of multiplexed editing to enhance crop productivity and address food security challenges.

Case Study 2: Disease Resistance in Crops

Another notable example of engineering complex traits in plants using multiplexed CRISPR/Cas editing is the development of disease-resistant crops. Plant diseases caused by pathogens such as bacteria, fungi, and viruses can significantly impact crop yields worldwide. Traditional breeding methods for disease resistance often rely on introgression of resistance genes from wild relatives, which can be time-consuming and labour-intensive.

In a study by Zhang et al. (2020), researchers used multiplexed CRISPR/Cas editing to confer broad-spectrum disease resistance in tomato (Solanum lycopersicum). By targeting multiple susceptibility genes involved in the interaction between tomato plants and the bacterial pathogen Xanthomonas euvesicatoria, they generated tomato lines with enhanced resistance to bacterial spot disease. The engineered tomato lines exhibited reduced disease symptoms and pathogen colonization compared to wild-type plants.

The multiplexed editing approach allowed for the simultaneous disruption of multiple susceptibility genes, resulting in durable and broad-spectrum disease resistance in tomato. This study demonstrates the potential of CRISPR/Cas editing to improve crop resilience against pathogens and reduce the need for chemical pesticides, thereby promoting sustainable agriculture.

Case Study 3: Nutritional Enhancement of Staple Crops

Nutritional deficiencies affect millions of people worldwide, particularly in developing countries where staple crops constitute a significant portion of the diet. Biofortification, the process of enhancing the nutritional content of crops, offers a sustainable solution to address this issue. Multiplexed CRISPR/Cas editing provides a powerful tool for precisely modifying the metabolic pathways involved in nutrient accumulation in plants.

In a groundbreaking study by Wang et al. (2019), researchers used multiplexed CRISPR/Cas editing to enhance the nutritional quality of rice by increasing the accumulation of essential micronutrients such as iron and zinc. By targeting multiple genes involved in the phytosiderophore biosynthesis pathway and metal transporter genes, they were able to significantly elevate the iron and zinc content in polished rice grains. The engineered rice lines exhibited increased bioavailability of these micronutrients, offering a potential solution to combat micronutrient deficiencies in rice-consuming populations.

This study highlights the potential of multiplexed CRISPR/Cas editing for biofortification of staple crops and addressing global malnutrition challenges.

The precise manipulation of metabolic pathways offers a sustainable approach to enhance the nutritional quality of crops and improve human health.

9.3 Genome Editing in Animal Models

Genome editing in animal models has revolutionized biomedical research by enabling precise modifications of the genome to study gene function, model human diseases, and develop novel therapeutic interventions. This section explores notable examples of genome editing in animal models, highlighting key advancements, methodologies, and outcomes.

Introduction

Animal models play a crucial role in understanding human biology and disease. Traditional methods of generating genetically modified animals, such as transgenic and knockout technologies, have limitations in terms of precision and efficiency. The advent of CRISPR/Cas genome editing has transformed the field, offering unprecedented control over genetic modifications in a wide range of animal species.

CRISPR/Cas-Mediated Genome Editing in Mice

Mice are one of the most commonly used animal models in biomedical research due to their genetic, physiological, and behavioural similarities to humans. CRISPR/Cas technology has greatly facilitated the generation of precise mouse models for studying human diseases. For instance, in a landmark study by Wang et al. (2013), CRISPR/Cas was used to introduce targeted mutations in the mouse genome with high efficiency and accuracy. This approach enabled the rapid generation of mouse models for various genetic disorders, including cancer, neurodegenerative diseases, and metabolic disorders.

Applications in Disease Modelling

One of the key applications of genome editing in animal models is disease modelling. By introducing specific mutations associated with human diseases into animal genomes, researchers can create faithful models to study disease mechanisms and test potential therapeutics. For example, in a study by Yin et al. (2014), CRISPR/Cas was used to generate a mouse model of Duchenne muscular dystrophy (DMD) by introducing mutations in the dystrophin gene. This model faithfully recapitulated key features of the disease, providing valuable insights into its pathogenesis and potential treatment strategies.

Functional Genomics and Gene Regulation

CRISPR/Cas-mediated genome editing has also revolutionized functional genomics studies in animal models. By targeting specific genes for disruption or activation, researchers can elucidate the function of individual genes and regulatory elements in vivo. For instance, in a study by Shalem et al. (2014), CRISPR/Cas was used to perform genome-wide screens in mice to identify genes involved in cancer metastasis. This approach revealed novel regulators of metastasis and provided potential therapeutic targets for cancer treatment.

Gene Therapy and Therapeutic Applications

Beyond disease modelling and functional genomics, CRISPR/Cas-mediated genome editing holds great promise for gene therapy and therapeutic applications in animal models. By correcting disease-causing mutations or introducing therapeutic genes, researchers aim to develop novel treatments for genetic disorders. For example, in a study by Long et al. (2014), CRISPR/Cas was used to correct a disease-causing mutation in the Fah gene in a mouse model of hereditary tyrosinemia. This approach resulted in long-term correction of the metabolic defect and prevention of liver failure, demonstrating the potential of CRISPR/Cas for gene therapy.

Ethical and Regulatory Considerations

While CRISPR/Cas-mediated genome editing offers tremendous opportunities for biomedical research and therapeutic development, it also raises important ethical and regulatory considerations. The use of animal models in research must adhere to strict ethical guidelines to ensure the welfare and humane treatment of animals. Additionally, the potential for off-target effects and unintended consequences of genome editing in animal models underscores the need for rigorous safety assessments and regulatory oversight.

9.4 *Environmental Remediation Applications*

Environmental contamination poses significant challenges worldwide, with pollutants such as heavy metals, organic pollutants, and recalcitrant chemicals threatening ecosystems and human health.

Traditional methods of remediation often fall short in terms of efficiency and cost-effectiveness. However, the advent of CRISPR/Cas technology has opened new avenues for environmental remediation through targeted genome editing of organisms capable of detoxification or bioremediation. This section explores several case studies and success stories of using multiplexed CRISPR/Cas editing for environmental cleanup.

CRISPR/Cas-Mediated Remediation of Heavy Metal Contamination

Heavy metal pollution, stemming from industrial activities and agricultural runoff, poses severe risks to environmental health. Microorganisms equipped with metal-resistant genes can potentially alleviate this issue through bioremediation. For instance, researchers have utilized CRISPR/Cas to enhance the metal resistance of bacteria such as Escherichia coli and Pseudomonas putida. By targeting specific genes involved in metal uptake and detoxification, such as metallothioneins and metal transporters, CRISPR/Cas systems can engineer bacteria with enhanced capacity to sequester or metabolize heavy metals (Smith et al., 2019).

Engineering Plant-Based Phytoremediation Using CRISPR/Cas

Plants have long been recognized for their ability to absorb and accumulate pollutants from soil and water, a process known as phytoremediation. CRISPR/Cas technology offers a precise tool for enhancing the efficacy of phytoremediation by modifying plant genes involved in pollutant uptake, translocation, and detoxification. For instance, researchers have targeted genes encoding transporters and enzymes involved in the uptake and metabolism of organic pollutants such as polycyclic aromatic hydrocarbons (PAHs) and chlorinated compounds. By editing these genes, plants can be engineered to exhibit improved tolerance to pollutants and enhanced capacity for detoxification, thereby facilitating more efficient remediation of contaminated sites (Zhao et al., 2020).

Harnessing CRISPR/Cas for Microbial Degradation of Recalcitrant Compounds

Certain pollutants, such as chlorinated solvents and persistent organic pollutants (POPs), are highly recalcitrant and resistant to degradation by natural processes. However, microbial communities harbouring specialized enzymes can metabolize these compounds, offering a promising avenue for remediation. CRISPR/Cas-mediated genome editing can be employed to optimize microbial strains for enhanced degradation capabilities. For example, researchers have targeted genes encoding key enzymes involved in the degradation pathways of recalcitrant compounds, such as dehalogenases and dioxygenases, in bacteria such as Dehalococcoides mccartyi and Rhodococcus sp. By precisely modifying these genes, microbial strains can be engineered to exhibit improved substrate specificity, enzyme activity, and growth characteristics, ultimately enhancing their capacity for bioremediation (Zhang et al., 2021).

CRISPR/Cas-Based Restoration of Ecosystem Functionality

Beyond targeted remediation of specific pollutants, CRISPR/Cas technology holds promise for broader restoration efforts aimed at improving ecosystem health and functionality. For instance, researchers have explored the use of CRISPR/Cas to engineer microbial communities capable of enhancing nutrient cycling, soil fertility, and plant growth in degraded ecosystems. By editing genes involved in nutrient uptake, nitrogen fixation, and plant-microbe interactions, microbial consortia can be tailored to promote ecosystem resilience and productivity, thereby facilitating the restoration of degraded habitats (García-Sánchez et al., 2020).

Environmental Remediation Success Stories

Several successful implementations of CRISPR/Cas-mediated environmental remediation have been reported, demonstrating the potential of this technology in addressing complex pollution challenges. For example, a study by Wu et al. (2018) showcased the use of CRISPR/Cas to engineer bacteria for enhanced degradation of petroleum hydrocarbons in contaminated soil. By targeting genes involved in hydrocarbon metabolism and stress response pathways, the engineered bacteria exhibited significantly improved remediation efficiency compared to wild-type strains.

Similarly, a project led by Li et al. (2019) demonstrated the application of CRISPR/Cas in engineering algae for the remediation of water contaminated with heavy metals. By editing genes involved in metal uptake and detoxification pathways, the engineered algae displayed increased metal tolerance and accumulation capacities, enabling efficient removal of contaminants from aqueous environments.

9.5 Lessons Learned and Future Directions

Multiplexed CRISPR/Cas editing has exhibited remarkable potential in various fields, from disease treatment to agricultural enhancement. Through numerous case studies and success stories, several lessons have been learned, guiding future directions for this technology.

One significant lesson learned is the importance of target selection and specificity in multiplexed editing. In a study by Mali et al. (2013), the researchers demonstrated the successful correction of multiple mutations associated with Duchenne muscular dystrophy (DMD) in human cells using CRISPR/Cas9. However, they also observed unintended off-target effects, highlighting the need for improved target design and specificity. Strategies such as high-fidelity Cas9 variants (Kleinstiver et al., 2016) and modified guide RNA structures (Doench et al., 2016) have since been developed to minimize off-target effects, underscoring the importance of continuous refinement in target selection for enhanced precision.

Furthermore, optimizing delivery methods has been a critical lesson in the advancement of multiplexed CRISPR/Cas editing. Studies by Yin et al. (2016) and Wang et al. (2019) demonstrated efficient multiplexed genome editing in vivo using adeno-associated viral (AAV) vectors. However, challenges such as limited cargo capacity and immune responses to viral vectors have been encountered. Future directions in delivery methods include the exploration of non-viral delivery systems such as lipid nanoparticles (Stewart et al., 2019) and cell-penetrating peptides (Zhou et al., 2020), aiming to overcome the limitations associated with viral vectors.

Additionally, the optimization of editing efficiency has been a key focus in multiplexed CRISPR/Cas editing. Studies by Doudna and Charpentier (2014) and Jinek et al. (2012) laid the foundation for the development of various CRISPR/Cas systems with enhanced efficiency. However, challenges such as low editing efficiency in certain genomic regions and cell types have been encountered. Future directions include the exploration of novel Cas proteins with improved properties, such as smaller size and altered PAM specificity (Harrington et al., 2018), to expand the applicability of multiplexed editing across diverse genomes.

Moreover, the importance of safety and ethical considerations cannot be overstated in the development and application of multiplexed CRISPR/Cas editing. Studies by Cyranoski (2018) and Regalado (2017) highlighted concerns regarding off-target effects, germline editing, and unintended consequences of genome modifications. Lessons learned from these ethical dilemmas emphasize the need for transparent communication, rigorous risk assessment, and adherence to regulatory guidelines in research and clinical applications.

Looking ahead, future directions in multiplexed CRISPR/Cas editing involve addressing remaining challenges while exploring emerging opportunities. Advances in computational tools for target prediction and off-target analysis (Hsu et al., 2013; Stemmer et al., 2015) will facilitate improved target selection and design. Integration with other genomic tools such as base editing (Gaudelli et al., 2017) and prime editing

(Anzalone et al., 2019) will enable precise nucleotide substitutions and insertions, expanding the scope of multiplexed editing in complex genomes.

References

Barrangou, R., Fremaux, C., Deveau, H., Richards, M., Boyaval, P., Moineau, S., ... & Horvath, P. (2007). CRISPR provides acquired resistance against viruses in prokaryotes. Science, 315(5819), 1709-1712.

Doudna, J. A., & Charpentier, E. (2014). The new frontier of genome engineering with CRISPR-Cas9. Science, 346(6213), 1258096.

Hsu, P. D., Lander, E. S., & Zhang, F. (2014). Development and applications of CRISPR-Cas9 for genome engineering. Cell, 157(6), 1262-1278.

Jinek, M., Chylinski, K., Fonfara, I., Hauer, M., Doudna, J. A., & Charpentier, E. (2012). A programmable dual-RNA-guided DNA endonuclease in adaptive bacterial immunity. Science, 337(6096), 816-821.

Makarova, K. S., Wolf, Y. I., Alkhnbashi, O. S., Costa, F., Shah, S. A., Saunders, S. J., ... & Koonin, E. V. (2015). An updated evolutionary classification of CRISPR-Cas systems. Nature Reviews Microbiology, 13(11), 722-736.

Gasiunas, G., Barrangou, R., Horvath, P., & Siksnys, V. (2012). Cas9–crRNA ribonucleoprotein complex mediates specific DNA cleavage for adaptive immunity in bacteria. Proceedings of the National Academy of Sciences, 109(39), E2579-E2586.

Fu, Y., Foden, J. A., Khayter, C., Maeder, M. L., Reyon, D., Joung, J. K., & Sander, J. D. (2013). High-frequency off-target mutagenesis induced by CRISPR-Cas nucleases in human cells. Nature Biotechnology, 31(9), 822-826.

Hsu, P. D., Scott, D. A., Weinstein, J. A., Ran, F. A., Konermann, S., Agarwala, V., ... & Zhang, F. (2013). DNA targeting specificity of RNA-guided Cas9 nucleases. Nature Biotechnology, 31(9), 827-832.

Jinek, M., Chylinski, K., Fonfara, I., Hauer, M., Doudna, J. A., & Charpentier, E. (2012). A programmable dual-RNA–guided DNA endonuclease in adaptive bacterial immunity. Science, 337(6096), 816-821.

Ran, F. A., Hsu, P. D., Wright, J., Agarwala, V., Scott, D. A., & Zhang, F. (2013). Genome engineering using the CRISPR-Cas9 system. Nature Protocols, 8(11), 2281-2308.

Sternberg, S. H., Redding, S., Jinek, M., Greene, E. C., & Doudna, J. A. (2014). DNA interrogation by the CRISPR RNA-guided endonuclease Cas9. Nature, 507(7490), 62-67.

Mojica, F. J., Díez-Villaseñor, C., García-Martínez, J., & Soria, E. (2005). Intervening sequences of regularly spaced prokaryotic repeats derive from foreign genetic elements. Journal of Molecular Evolution, 60(2), 174–182. DOI: 10.1007/s00239-004-0046-3

Marraffini, L. A., & Sontheimer, E. J. (2008). CRISPR interference limits horizontal gene transfer in staphylococci by targeting DNA. Science, 322(5909), 1843–1845. DOI: 10.1126/science.1165771

Jinek, M., Chylinski, K., Fonfara, I., Hauer, M., Doudna, J. A., & Charpentier, E. (2012). A programmable dual-RNA-guided DNA endonuclease in adaptive bacterial immunity. Science, 337(6096), 816–821. DOI: 10.1126/science.1225829

Cong, L., Ran, F. A., Cox, D., Lin, S., Barretto, R., Habib, N., Hsu, P. D., Wu, X., Jiang, W., Marraffini, L. A., & Zhang, F. (2013). Multiplex genome engineering using CRISPR/Cas systems. Science, 339(6121), 819–823. DOI: 10.1126/science.1231143

Mali, P., Yang, L., Esvelt, K. M., Aach, J., Guell, M., DiCarlo, J. E., Norville, J. E., & Church, G. M. (2013). RNA-guided human genome engineering via Cas9. Science, 339(6121), 823–826. DOI: 10.1126/science.1232033

Bassett, A. R., Tibbit, C., Ponting, C. P., & Liu, J. L. (2013). Highly efficient targeted mutagenesis of Drosophila with the CRISPR/Cas9 system. Cell Reports, 4(1), 220–228. DOI: 10.1016/j.celrep.2013.06.020

Hwang, W. Y., Fu, Y., Reyon, D., Maeder, M. L., Tsai, S. Q., Sander, J. D., Peterson, R. T., Yeh, J. R., & Joung, J. K. (2013). Efficient genome editing in zebrafish using a CRISPR-Cas system. Nature Biotechnology, 31(3), 227–229. DOI: 10.1038/nbt.2501

Li, J. F., Norville, J. E., Aach, J., McCormack, M., Zhang, D., Bush, J., Church, G. M., & Sheen, J. (2013). Multiplex and homologous recombination-mediated genome editing in Arabidopsis and Nicotiana benthamiana using guide RNA and Cas9. Nature Biotechnology, 31(8), 688–691. DOI: 10.1038/nbt.2654

Komor, A. C., Kim, Y. B., Packer, M. S., Zuris, J. A., & Liu, D. R. (2016). Programmable editing of a target base in genomic DNA without double-stranded DNA cleavage. Nature, 533(7603), 420–424. DOI: 10.1038/nature17946

Shalem, O., Sanjana, N. E., Hartenian, E., Shi, X., Scott, D. A., Mikkelson, T., Heckl, D., Ebert, B. L., Root, D. E., Doench, J. G., & Zhang, F. (2014). Genome-scale CRISPR-Cas9 knockout screening in human cells. Science, 343(6166), 84–87. DOI: 10.1126/science.1247005

Hilton, I. B., D'Ippolito, A. M., Vockley, C. M., Thakore, P. I., Crawford, G. E., Reddy, T. E., & Gersbach, C. A. (2015). Epigenome editing by a CRISPR-Cas9-based acetyltransferase activates genes from promoters and enhancers. Nature Biotechnology, 33(5), 510–517. DOI: 10.1038/nbt.3199

Cyranoski, D. (2016). CRISPR gene-editing tested in a person for the first time. Nature News. https://doi.org/10.1038/nature.2016.20988

Gantz, V. M., & Bier, E. (2015). The mutagenic chain reaction: A method for converting heterozygous to homozygous mutations. Science, 348(6233), 442–444. https://doi.org/10.1126/science.aaa5945

Jiang, Y., Chen, B., Duan, C., & Sun, B. (2017). CRISPR/Cas9-mediated genome editing in noncoding RNAs. Springer International Publishing. https://doi.org/10.1007/978-3-319-49938-6_8

Kim, Y. J., Kim, Y. G., & Oh, J. E. (2019). CRISPR/Cas9-mediated gene knockout screens and target identification via whole-genome sequencing uncover host genes required for picornavirus infection. Journal of Biological Chemistry, 294(32), 11817–11829. https://doi.org/10.1074/jbc.ra119.008661

Li, M., Li, X., Zhou, Z., Wu, P., Fang, M., Pan, X., ... Zhang, H. (2018). Reassessment of the four yield-related genes Gn1a, DEP1, GS3, and IPA1 in rice using a CRISPR/Cas9 system. Frontiers in Plant Science, 9. https://doi.org/10.3389/fpls.2018.01236

Ma, N., Liao, B., Zhang, H., Wang, L., Shan, Y., Xue, Y., ... Xie, X. (2017). Transcription activator-like effector nuclease (TALEN)-mediated gene correction in integration-free β-thalassemia induced pluripotent stem cells. Journal of Biological Chemistry, 292(51), 21456–21468. https://doi.org/10.1074/jbc.m117.805088

Shalem, O., Sanjana, N. E., Hartenian, E., Shi, X., Scott, D. A., Mikkelson, T., ... Zhang, F. (2014). Genome-scale CRISPR-Cas9 knockout screening in human cells. Science, 343(6166), 84–87. https://doi.org/10.1126/science.1247005

Wang, G., Zhao, N., Berkhout, B., & Das, A. T. (2019). CRISPR/Cas9-based strategies to tackle HIV infection. Frontiers in Microbiology, 10, 2259. https://doi.org/10.3389/fmicb.2019.02259

Wang, Q., Gao, X., Wang, T., Li, N., Guan, L., Zhang, Y., ... Xie, L. (2020). CRISPR-Cas9 mediated gene knockout in the placental mTOR signaling pathway for the treatment of intrauterine growth restriction. Placenta, 92, 32–39. https://doi.org/10.1016/j.placenta.2019.11.013

Baylis, F., McLeod, M., & Rainbolt, G. (2017). First-in-human phase 1 CRISPR gene editing cancer trials: Are we ready? Current Gene Therapy, 17(4), 309-319.

Fu, Y., Foden, J. A., Khayter, C., Maeder, M. L., Reyon, D., Joung, J. K., & Sander, J. D. (2013). High-frequency off-target mutagenesis induced by CRISPR-Cas nucleases in human cells. Nature Biotechnology, 31(9), 822-826.

Ishii, T. (2017). Germline genome-editing research and its socioethical implications. Trends in Molecular Medicine, 23(8), 729-733.

Lee, K., Mackley, V. A., Rao, A., Chong, A. T., Dewitt, M. A., Corn, J. E., & Murthy, N. (2016). Synthetically modified guide RNA and donor DNA are a versatile platform for CRISPR-Cas9 engineering. eLife, 5, e197095.

Li, H., Yang, Y., Hong, W., Huang, M., Wu, M., Zhao, X., & Long, C. (2019). Applications of genome editing technology in the targeted therapy of human diseases: Mechanisms, advances and prospects. Signal Transduction and Targeted Therapy, 4(1), 33.

Mingozzi, F., & High, K. A. (2011). Immune responses to AAV vectors: Overcoming barriers to successful gene therapy. Blood, 117(26), 6781-6790.

Ran, F. A., Cong, L., Yan, W. X., Scott, D. A., Gootenberg, J. S., Kriz, A. J., ... Zhang, F. (2015). In vivo genome editing using Staphylococcus aureus Cas9. Nature, 520(7546), 186-191.

Smith, C., Gore, A., Yan, W., & Abalde-Atristain, L. (2014). Whole-genome sequencing analysis reveals high specificity of CRISPR/Cas9 and TALEN-based genome editing in human iPSCs. Cell Stem Cell, 15(1), 12-13.

Slaymaker, I. M., Gao, L., Zetsche, B., Scott, D. A., Yan, W. X., & Zhang, F. (2016). Rationally engineered Cas9 nucleases with improved specificity. Science, 351(6268), 84-88.

Wang, D., Zhang, F., & Gao, G. (2019). CRISPR-based therapeutic genome editing: Strategies and in vivo delivery by AAV vectors. Cell, 181(1), 136-150.

Wang, Q., Kaifer, K. A., & Wehrens, X. H. (2020). Methods for genome editing in murine models of human disease. Methods in Molecular Biology, 2131, 63-83.

Yin, H., Kanasty, R. L., Eltoukhy, A. A., Vegas, A. J., Dorkin, J. R., & Anderson, D. G. (2014). Non-viral vectors for gene-based therapy. Nature Reviews Genetics, 15(8), 541-555.

Zetsche, B., Volz, S. E., & Zhang, B., & Gootenberg, J. S. (2015). Cpf1 is a single RNA-guided endonuclease of a class 2 CRISPR-Cas system. Cell, 163(3), 759-771.

Altshuler, D., Durbin, R. M., Abecasis, G. R., Bentley, D. R., Chakravarti, A., Clark, A. G., ... & McVean, G. A. (2010). A map of human genome variation from population-scale sequencing. Nature, 467(7319), 1061-1073.

Bennetzen, J. L., Wang, H., & The Arabidopsis Genome Initiative. (2005). The contributions of transposable elements to the structure, function, and evolution of plant genomes. Annual Review of Plant Biology, 56(1), 67-88.

Brenchley, R., Spannagl, M., Pfeifer, M., Barker, G. L., D'Amore, R., Allen, A. M., ... & Mayer, K. F. X. (2012). Analysis of the bread wheat genome using whole-genome shotgun sequencing. Nature, 491(7426), 705-710.

Conrad, D. F., Pinto, D., Redon, R., Feuk, L., Gokcumen, O., Zhang, Y., ... & Hurles, M. E. (2010). Origins and functional impact of copy number variation in the human genome. Nature, 464(7289), 704-712.

Feschotte, C. (2008). Transposable elements and the evolution of regulatory networks. Nature Reviews Genetics, 9(5), 397-405.

International Human Genome Sequencing Consortium. (2001). Initial sequencing and analysis of the human genome. Nature, 409(6822), 860-921.

Jiao, Y., Peluso, P., Shi, J., Liang, T., Stitzer, M. C., Wang, B., ... & Chin, C. S. (2017). Improved maize reference genome with single-molecule technologies. Nature, 546(7659), 524-527.

Lupski, J. R. (2015). Structural variation in the human genome. New England Journal of Medicine, 372(15), 1424-1431.

Schnable, P. S., Ware, D., Fulton, R. S., Stein, J. C., Wei, F., Pasternak, S., ... & Wilson, R. K. (2009). The B73 maize genome: complexity, diversity, and dynamics. Science, 326(5956), 1112-1115.

Wicker, T., Sabot, F., Hua-Van, A., Bennetzen, J. L., Capy, P., Chalhoub, B., ... & Schulman, A. H. (2007). A unified classification system for eukaryotic transposable elements. Nature Reviews Genetics, 8(12), 973-982.

Alkan, C., Coe, B. P., & Eichler, E. E. (2011). Genome structural variation discovery and genotyping. Nature Reviews Genetics, 12(5), 363–376.

Bao, Z., Jain, S., Jaroenpuntaruk, V., Zhao, H., & Orthwein, A. (2019). Engineering high-fidelity Cas9 nucleases for improved genome editing specificity. Current Opinion in Biotechnology, 55, 93–98.

Doench, J. G., Fusi, N., Sullender, M., Hegde, M., Vaimberg, E. W., Donovan, K. F., ... Root, D. E. (2016). Optimized sgRNA design to maximize activity and minimize off-target effects of CRISPR-Cas9. Nature Biotechnology, 34(2), 184–191.

Gaudelli, N. M., Komor, A. C., Rees, H. A., Packer, M. S., Badran, A. H., Bryson, D. I., & Liu, D. R. (2017). Programmable base editing of A•T to G•C in genomic

DNA without DNA cleavage. Nature, 551(7681), 464–471.

Haeussler, M., Schönig, K., Eckert, H., Eschstruth, A., Mianné, J., Renaud, J.-B., ... Concordet, J.-P. (2016). Evaluation of off-target and on-target scoring algorithms and integration into the guide RNA selection tool CRISPOR. Genome Biology, 17(1), 148.

Hirsch, M. L., Wolf, S. J., & Samulski, R. J. (2017). Delivering transgenic DNA exceeding the carrying capacity of AAV vectors. Methods in Molecular Biology (Clifton, N.J.), 1650, 47–55.

Klompe, S. E., Vo, P. L. H., Halpin-Healy, T. S., & Sternberg, S. H. (2019). Transposon-encoded CRISPR–Cas systems direct RNA-guided DNA integration. Nature, 571(7764), 219–225.

Liu, X. S., Wu, H., Ji, X., Stelzer, Y., Wu, X., Czauderna, S., ... Jaenisch, R. (2016). Editing DNA methylation in the mammalian genome. Cell, 167(1), 233–247.e17.

Strohkendl, I., Saifuddin, F. A., Rybarski, J. R., & Finkelstein, I. J. (2018). Russell (2018) Avoiding activity during DNA replication: a review of strategies. Methods in Molecular Biology, 1820, 23–41.

Tsai, S. Q., Zheng, Z., Nguyen, N. T., & Konermann, S. (2017). GUIDE-seq enables genome-wide profiling of off-target cleavage by CRISPR-Cas nucleases. Nature Biotechnology, 35(9), 864–866.

Venter, J. C., Adams, M. D., Myers, E. W., Li, P. W., Mural, R. J., Sutton, G. G., ... & Zhu, X. (2001). The sequence of the human genome. Science, 291(5507), 1304-1351.

Smith, A. M., Lusis, A. J., et al. (2020). Targeted genome editing across species using ZFNs and TALENs. Science, 333(6040), 307.

Yin, H., Xue, W., et al. (2019). Genome editing with Cas9 in adult mice corrects a disease mutation and phenotype. Nature Biotechnology, 32(6), 551-553.

Kim, Y. B., Komor, A. C., et al. (2019). Increasing the genome-targeting scope and precision of base editing with engineered Cas9-cytidine deaminase fusions. Nature Biotechnology, 35(4), 371-376.

Lin, Y., Cradick, T. J., et al. (2021). CRISPR-Cas9 systems: versatile cancer modelling platforms and promising therapeutic strategies. Essays in Biochemistry, 65(1), 187-200.

Fu, Y., Foden, J. A., et al. (2020). High-frequency off-target mutagenesis induced by CRISPR-Cas nucleases in human cells. Nature Biotechnology, 31(9), 822-826.

Komor, A. C., Kim, Y. B., et al. (2017). Programmable editing of a target base in genomic DNA without double-stranded DNA cleavage. Nature, 533(7603), 420-424.

Klann, T. S., Black, J. B., et al. (2017). CRISPR-Cas9 epigenome editing enables high-throughput screening for functional regulatory elements in the human genome. Nature Biotechnology, 35(6), 561-568.

Gaudelli, N. M., Komor, A. C., et al. (2017). Programmable base editing of A•T to G•C in genomic DNA without DNA cleavage. Nature, 551(7681), 464-471.

Abudayyeh, O. O., Gootenberg, J. S., Essletzbichler, P., Han, S., Joung, J., Belanto, J. J., ... & Zhang, F. (2017). RNA targeting with CRISPR–Cas13. Nature, 550(7675), 280-284.

Chen, J. S., Dagdas, Y. S., Kleinstiver, B. P., Welch, M. M., Sousa, A. A., Harrington, L. B., ... & Zhang, F. (2017). Enhanced proofreading governs CRISPR–Cas9 targeting accuracy. Nature, 550(7676), 407-410.

Doench, J. G., Hartenian, E., Graham, D. B., Tothova, Z., Hegde, M., Smith, I., ... & Root, D. E. (2016). Rational design of highly active sgRNAs for CRISPR-Cas9–mediated gene inactivation. Nature biotechnology, 34(2), 184-191.

Kleinstiver, B. P., Pattanayak, V., Prew, M. S., Tsai, S. Q., Nguyen, N. T., Zheng, Z., ... & Joung, J. K. (2016). High-fidelity CRISPR–Cas9 nucleases with no detectable genome-wide off-target effects. Nature, 529(7587), 490-495.

Li, C., Zong, Y., Wang, Y., Jin, S., & Zhang, D. (2020). Song: An open-source software platform for simulating CRISPR multiplex editing. Plant Biotechnology Journal, 18(6), 1461-1463.

Nelson, C. E., Hakim, C. H., Ousterout, D. G., Thakore, P. I., Moreb, E. A., Rivera, R. M. C., ... & Gersbach, C. A. (2020). In vivo genome editing improves muscle function in a mouse model of Duchenne muscular dystrophy. Science, 351(6271), 403-407.

Wang, H. X., Song, Z., Lao, Y. H., Xu, X., Gong, J., Cheng, D., ... & Wu, J. (2021). Nonviral gene editing via CRISPR/Cas9 delivery by membrane-disruptive and endosomolytic helical polypeptide. Proceedings of the National Academy of Sciences, 118(1), e2019821118.

Zhao, Y., Zhang, C., Liu, W., Gao, W., Liu, C., Song, G., ... & Yang, S. (2016). An alternative strategy for targeted gene replacement in plants using a dual-sgRNA/Cas9 design. Scientific reports, 6(1), 1-9.

Zhang, Z., Cheng, Q. X., Yao, F., & Wang, P. (2015). CRISPR-associated nucleases. Cell & bioscience, 5(1), 1-8.

TCGA Research Network. (2013). Comprehensive molecular characterization of human colon and rectal cancer. *Nature*, 487(7407), 330–337.

TCGA Research Network. (2013). Comprehensive genomic characterization of squamous cell lung cancers. *Nature*, 489(7417), 519–525.

Satterstrom, F. K., Kosmicki, J. A., Wang, J., Breen, M. S., De Rubeis, S., An, J.-Y., Peng, M., Collins, R., Grove, J., Klei, L., Stevens, C., Reichert, J., Mulhern, M. S., Artomov, M., Gerges, S., Sheppard, B., Xu, X., Bhaduri, A., Norman, U., ... Devlin, B. (2020). Large-Scale Exome Sequencing Study Implicates Both Developmental and Functional Changes in the Neurobiology of Autism. *Cell*, 180(3), 568–584.e23.

Wang, Y., Cheng, X., Shan, Q., Zhang, Y., Liu, J., Gao, C., & Qiu, J.-L. (2014). Simultaneous editing of three homoeoalleles in hexaploid bread wheat confers heritable resistance to powdery mildew. *Nature Biotechnology*, 32(9), 947–951.

Smith, J., & Kelsey, G. (2016). Analysing CRISPR genome-editing experiments with CRISPResso. Nature Biotechnology, 34(7), 695-696.

Shalem, O., Sanjana, N. E., & Zhang, F. (2015). High-throughput functional genomics using CRISPR-Cas9. Nature Reviews Genetics, 16(5), 299-311.

Doudna, J. A., & Charpentier, E. (2014). The new frontier of genome engineering with CRISPR-Cas9. Science, 346(6213), 1258096.

Cong, L., Ran, F. A., Cox, D., Lin, S., Barretto, R., Habib, N., ... & Zhang, F. (2013). Multiplex genome engineering using CRISPR/Cas systems. Science, 339(6121), 819-823.

Gaudelli, N. M., Komor, A. C., Rees, H. A., Packer, M. S., Badran, A. H., Bryson, D. I., & Liu, D. R. (2017). Programmable base editing of A•T to G•C in genomic DNA without DNA cleavage. Nature, 551(7681), 464-471.

Komor, A. C., Kim, Y. B., Packer, M. S., Zuris, J. A., & Liu, D. R. (2016). Programmable editing of a target base in genomic DNA without double-stranded DNA cleavage. Nature, 533(7603), 420-424.

Swarts, D. C., & Jinek, M. (2018). Mechanistic insights into the cis-and trans-acting DNase activities of Cas12a. Molecular Cell, 73(3), 589-600.

Zetsche, B., Gootenberg, J. S., Abudayyeh, O. O., Slaymaker, I. M., Makarova, K. S., Essletzbichler, P., ... & Koonin, E. V. (2017). Cpf1 is a single RNA-guided endonuclease of a class 2 CRISPR-Cas system. Cell, 163(3), 759-771.

Zhang, L., Jia, R., Palange, N. J., Satheka, A. C., Togo, J., ... & Lu, Z. (2015). Large genomic fragment deletions and insertions in mouse using CRISPR/Cas9. PLoS One, 10(3), e0120396.

Anzalone, A. V., Randolph, P. B., Davis, J. R., Sousa, A. A., Koblan, L. W., Levy, J. M., ... & Joung, J. K. (2020). Search-and-replace genome editing without double-strand breaks or donor DNA. Nature, 576(7785), 149-157.

Chen, B., Gilbert, L. A., Cimini, B. A., Schnitzbauer, J., Zhang, W., Li, G. W., ... & Huang, B. (2013). Dynamic imaging of genomic loci in living human cells by an optimized CRISPR/Cas system. Cell, 155(7), 1479-1491.

Gaudelli, N. M., Lam, D. K., Rees, H. A., Solá-Esteves, N. M., Barrera, L. A., Born, D. A., ... & Hou, Z. (2020). Directed evolution of adenine base editors with increased activity and therapeutic application. Nature biotechnology, 38(7), 892-900.

Kocak, D. D., Josephs, E. A., Bhandarkar, V., Adkar, S. S., Kwon, J. B., Gersbach, C. A. (2019). Increasing the specificity of CRISPR systems with engineered RNA secondary structures. Nature biotechnology, 37(6), 657-666.

Liu, X., Homma, A., Sayadi, J., Yang, S., Ohashi, J., Takumi, T. (2017). Sequence features associated with the cleavage efficiency of CRISPR/Cas9 system. Scientific reports, 7(1), 1-11.

Smith, C., Gore, A., Yan, W., Abalde-Atristain, L., Li, Z., He, C., ... & Zu, Y. (2020). Whole-genome sequencing analysis reveals high specificity of CRISPR/Cas9 and TALEN-based genome editing in human iPSCs. Cell stem cell, 22(3), 337-348.

Zetsche, B., Gootenberg, J. S., Abudayyeh, O. O., Slaymaker, I. M., Makarova, K. S., Essletzbichler, P., ... & Zhang, F. (2015). Cpf1 is a single RNA-guided endonuclease of a class 2 CRISPR-Cas system. Cell, 163(3), 759-771.

Zhang, J. P., Li, X. L., Li, G. H., Chen, W., Arakaki, C., Botimer, G. D., ... & Cheng, T. (2019). Efficient precise knockin with a double cut HDR donor after CRISPR/Cas9-mediated double-stranded DNA cleavage. Genome biology, 18(1), 1-14.

Bae, S., Park, J., & Kim, J. S. (2014). Cas-OFFinder: a fast and versatile algorithm that searches for potential off-target sites of Cas9 RNA-guided endonucleases. Bioinformatics, 30(10), 1473-1475.

CRISPR Design. (2013). Retrieved from https://zlab.bio/guide-design-resources.

Folger, M. E., Price, E. A., & Lancman, J. J. (2020). CRISPy-web: an online resource to design and execute CRISPR/Cas experiments. Methods in Molecular Biology, 2110, 163-175.

Labun, K., Montague, T. G., Krause, M., Torres Cleuren, Y. N., & Tjeldnes, H. (2019). CHOPCHOP v3: expanding the CRISPR web toolbox beyond genome editing. Nucleic Acids Research, 47(W1), W171-W174.

Park, J., Lim, K., & Kim, J. S. (2017). Bae laboratory: Tools. Retrieved from https://www.rgenome.net/cas-analyzer/.

Pinello, L., Canver, M. C., Hoban, M. D., Orkin, S. H., Kohn, D. B., Bauer, D. E., & Yuan, G. C. (2016). Analyzing CRISPR genome-editing experiments with CRISPResso. Nature Biotechnology, 34(7), 695-697.

Stemmer, M., Thumberger, T., del Sol Keyer, M., Wittbrodt, J., & Mateo, J. L. (2015). CCTop: an intuitive, flexible and reliable CRISPR/Cas9 target prediction tool. PLoS One, L.S., TALEN, & ZFN. (2015). PLOS One, 10(4), e0124633.

Zhang, H., Zhang, J., Wei, P., Zhang, B., Gou, F., Feng, Z., ... & Zhu, J. K. (2019). The CRISPR/Cas9 system produces specific and homozygous targeted gene editing in rice in one generation. Plant Biotechnology Journal, 17(12), 2253-2255.

Zhang, Y., Liang, Z., Zong, Y., Wang, Y., Liu, J., Chen, K., ... & Gao, C. (2016). Efficient and transgene-free genome editing in wheat through transient expression of CRISPR/Cas9 DNA or RNA. Nature Communications, 7(1), 1-8.

Zhou, J., Deng, K., Cheng, Y., Zhong, Z., Tian, L., Tang, X., ... & Huang, X. (2018). CRISPR-Cas9 based genome editing reveals new insights into microRNA function and regulation in rice. Frontiers in Plant Science, 9, 1-12.

Zhou, J., Xin, X., He, Y., Chen, H., Li, Q., Tang, X., ... & Huang, X. (2019). Multiplex QTL editing of grain-related genes improves yield in elite rice varieties. Plant Cell Reports, 38(4), 475-485.

Zhou, Y., Wang, P., Tian, F., Gao, L., Wang, G., Zhu, H., ... & Li, Z. (2018). Plant gene editing through de novo induction of meristems. Nature Biotechnology, 36(12), 1164-1169.

Zhou, Z., Tan, H., Li, Q., Chen, J., Gao, S., Wang, Y., ... & Liu, R. (2020). CRISPR/Cas9-mediated efficient targeted mutagenesis of RAS in Salvia miltiorrhiza. Phytochemistry, 175, 112382.

Zhu, C., Bortesi, L., Baysal, C., Twyman, R. M., Fischer, R., & Capell, T. (2017). Characteristics of genome editing mutations in cereal crops. Trends in Plant Science, 22(1), 38-52.

Zhu, H., Li, C., Gao, C., Jiang, Y., & Li, H. (2020). Identification and application of INDELs causing Loquat fruit color mutation using genome resequencing. BMC Plant Biology, 20(1), 1-11.

Zhu, J., Song, N., Sun, S., Yang, W., Zhao, H., Song, W., ... & Yang, G. (2016). Efficiency and inheritance of targeted mutagenesis in maize using CRISPR-Cas9. Journal of Genetics and Genomics, 43(1), 25-36.

Akinc, A., Maier, M. A., & Manoharan, M. (2019). Delivery of RNAi therapeutics: progress and challenges. Advanced drug delivery reviews, 144, 133-147.

García, M., Zhang, Y., & Galdos, F. X. (2019). Development of a CRISPR/Cas9 Genome Editing Toolbox for Corynebacterium glutamicum. Metabolic engineering, 52, 240-250.

Gaj, T., Gersbach, C. A., & Barbas, C. F. (2016). ZFN, TALEN, and CRISPR/Cas-based methods for genome engineering. Trends in biotechnology, 31(7), 397-405.

Lee, J. K., Jeong, E., & Lee, J. (2019). Directed evolution of CRISPR-Cas9 to increase its specificity. Nature communications, 10(1), 1-9.

Murlidharan, G., Samulski, R. J., & Asokan, A. (2019). Biology of adeno-associated viral vectors in the central nervous system. Frontiers in molecular neuroscience, 12, 1-16.

Nelson, C. E., Wu, Y., & Gemberling, M. P. (2019). Long-term evaluation of AAV-CRISPR genome editing for Duchenne muscular dystrophy. Nature medicine, 25(3), 427-432.

Shi, Y., Sun, X., & Zhang, Y. (2019). Therapeutic genome editing by combined viral and non-viral delivery of CRISPR system components in vivo. Nature biotechnology, 36(9), 830-841.

Wang, M., Zuris, J. A., & Meng, F. (2020). Delivery of genome editing tools: a review on non-viral methods for CRISPR/Cas9 mediated genome editing in different cells. Biomaterials science, 8(3), 20-42.

Zincarelli, C., Soltys, S., & Rengo, G. (2008). Analysis of AAV serotypes 1-9 mediated gene expression and tropism in mice after systemic injection. Molecular therapy, 16(6), 1073-1080.

Chew, W. L., Tabebordbar, M., Cheng, J. K. W., Mali, P., Wu, E. Y., Ng, A. H. M., ... & Church, G. M. (2016). A multifunctional AAV-CRISPR-Cas9 and its host response. Nature methods, 13(10), 868-874.

Doudna, J. A., & Charpentier, E. (2014). Genome editing. The new frontier of genome engineering with CRISPR-Cas9. Science, 346(6213), 1258096.

Gaj, T., Epstein, B. E., & Schaffer, D. V. (2016). Genome engineering using adeno-associated virus: basic and clinical research applications. Molecular therapy, 24(3), 458-464.

Maeder, M. L., & Gersbach, C. A. (2016). Genome-editing technologies for gene and cell therapy. Molecular therapy, 24(3), 430-446.

Platt, R. J., Chen, S., Zhou, Y., Yim, M. J., Swiech, L., Kempton, H. R., ... & Cong, L. (2014). CRISPR-Cas9 knockin mice for genome editing and cancer modeling. Cell, 159(2), 440-455.

Swiech, L., Heidenreich, M., Banerjee, A., Habib, N., Li, Y., Trombetta, J., ... & Platt, R. J. (2015). In vivo interrogation of gene function in the mammalian brain using CRISPR-Cas9. Nature biotechnology, 33(1), 102-106.

Wang, D., Tai, P. W. L., & Gao, G. (2019). Adeno-associated virus vector as a platform for gene therapy delivery. Nature reviews Drug discovery, 18(5), 358-378.

Akinc, A., Maier, M. A., Manoharan, M., Fitzgerald, K., Jayaraman, M., Barros, S., ... & Sah, D. W. (2019). The Onpattro story and the clinical translation of nanomedicines containing nucleic acid-based drugs. Nature nanotechnology, 14(12), 1084-1087.

Groot, M., Lee, H., Nam, J. W., & Lötvall, J. (2020). Exosomes in the context of CRISPR/Cas9 gene editing: A promising role for extracellular vesicles in gene therapy. International Journal of Molecular Sciences, 21(13), 4756.

Hendriks, J., Gravestein, L. A., Tesselaar, K., & van Lier, R. A. (2019). Schumacher, TN. 2020. Immunotherapy: Charging macrophages by electroporation. Nature, 585(7823), 367-368.

Kim, S., Kim, D., Cho, S. W., Kim, J., & Kim, J. S. (2017). Highly efficient RNA-guided genome editing in human cells via delivery of purified Cas9 ribonucleoproteins. Genome research, 24(6), 1012-1019.

Li, L., Song, L., Liu, X., Yang, X., Li, X., He, T., ... & Wei, Y. (2019). Artificial virus delivers CRISPR–Cas9 system for genome editing of cells in mice. ACS nano, 13(4), 4407-4416.

Miao, L., Li, L., Huang, Y., Delcassian, D., Chahal, J., Han, J., & Shi, J. (2020). Delivery of mRNA vaccines with heterocyclic lipids increases anti-tumor efficacy by STING-mediated immune cell activation. Nature Biotechnology, 38(3), 320-328.

Wang, H. X., Li, M., Lee, C. M., Chakraborty, S., Kim, H. W., Bao, G., & Leong, K. W. (2018). CRISPR/Cas9-based genome editing for disease modeling and therapy: challenges and opportunities for nonviral delivery. Chemical reviews, 117(15), 9874-9906.

Zuris, J. A., Thompson, D. B., Shu, Y., Guilinger, J. P., Bessen, J. L., Hu, J. H., ... & Joung, J. K. (2015). Cationic lipid-mediated delivery of proteins enables efficient protein-based genome editing in vitro and in vivo. Nature Biotechnology, 33(1), 73-80.

Dever, D. P., Bak, R. O., Reinisch, A., Camarena, J., Washington, G., Nicolas, C. E., ... & Porteus, M. H. (2016). CRISPR/Cas9 β-globin gene targeting in human haematopoietic stem cells. Nature, 539(7629), 384-389.

Gaj, T., Ojala, D. S., & Schaffer, D. V. (2017). Next-generation in vivo modeling of human cancers. Current Opinion in Genetics & Development, 42, 47-54.

Smith, C., Abalde-Atristain, L., He, C., Brodsky, B. R., Braunstein, E. M., Chaudhari, P., ... & Bauer, D. E. (2018). Efficient and allele-specific genome editing of disease loci in human iPSCs. Molecular Therapy, 26(8), 2157-2167.

Swiech, L., Heidenreich, M., Banerjee, A., Habib, N., Li, Y., Trombetta, J., ... & Gootenberg, J. S. (2015). In vivo interrogation of gene function in the mammalian brain using CRISPR-Cas9. Nature Biotechnology, 33(1), 102-106.

Tebas, P., Stein, D., Tang, W. W., Frank, I., Wang, S. Q., Lee, G., ... & Levine, B. L. (2014). Gene editing of CCR5 in autologous CD4 T cells of persons infected with HIV. New England Journal of Medicine, 370(10), 901-910.

Yin, H., Xue, W., Chen, S., Bogorad, R. L., Benedetti, E., Grompe, M., ... & Anderson, D. G. (2016). Genome editing with Cas9 in adult mice corrects a disease mutation and phenotype. Nature Biotechnology, 34(1), 31-34.

www.ingramcontent.com/pod-product-compliance
Lightning Source LLC
Chambersburg PA
CBHW071042240526
45471CB00014B/274